INVESTIGATING EVOLUTIONARY BIOLOGY IN THE LABORATORY

WILLIAM F. McCOMAS, Editor
University of Southern California
Los Angeles, California

Published by the National Association of Biology Teachers (NABT)
11250 Roger Bacon Drive #19, Reston, Virginia 22090

ISBN 0-941212-15-7

Cover photo courtesy of William F. McComas, Assistant Professor of Science Education at the University of Southern California, School of Education, Los Angeles, California 90089-0031. The image of one of "Darwin's" finches riding on the back of a giant tortoise was taken on Santa Cruz, one of the Galapagos Islands.

Printed in the United States of America by Lancaster Press, Lancaster, Pennsylvania

TABLE OF CONTENTS

I. Introduction

- Evolution Education in the Laboratory, *W.F. McComas* 7
- Misconceptions in Evolution Education, *A.I. Colburn* 11
- Reducing Resistance to Evolution Education, *M.P. Clough* 15

II. Evidences of Evolution

- Biochemical Evidence for Evolution, *D.J. Dyman* 23
- Evidences of Evolution Through Comparative Anatomy, *R.A. Coler* 27

III. General Evolutionary Principles

- How Long Is a Long Time? *W.F. McComas* 31
- Illustrating the Principle of the Filter Bridge, *R.E. Collins* and *R.W. Olsen* 40

IV. Variation Within the Species

- Illustrating Variation and Adaptation at the Zoo, *W.F. McComas* 45
- A Species Approach to Evolution Education, *D.B. Rosenthal* 51
- Demonstrating Variation within the Species, *D.H. Keown* 54

V. Biotic Potential and Survival

- The Arithmetic of Evolution, *R. Esterle* 59
- Demonstrating Biotic Potential, *D.H. Keown* 63
- Simulating Population Dynamics, *B.J. Alters* 64

VI. Adaptation

- Modeling the Principles of Adaptive Radiation, *L. Blackbeer et al.* 69
- The Birds and the Beaks, *R. Esterle* 74
- Demonstrating Color Adaptation in Food Selection, *J. Abruscato* and *L. Kenney* 77

VII. Simulating Natural Selection

- Simulating Evolution, *R.C. Stebbins* and *B. Allen* 81
- Simulating Natural Selection, *R. Patterson et al.* 88
- Demonstrating Natural Selection Through the Survival Value of
  Cryptic Coloration and Apostatic Selection, *J.A. Allen et al.* 91
- Demonstrating the Effects of Selection, *J.E. Thomerson* 98
- A Model of Microevolution in Action, *L.A. Welch* 102

VIII. Proposing Phylogenies

- Using Alumenontos to Investigate Evolution and Phylogeny, *S.J. Hageman* 111
- A Simulation Model Approach to the Study of Evolution, *J.A. Dawes* 117
- Illustrating Phylogeny and Classification, *J.M. Burns* 120
- The Caminalcule Family Tree, *D.J. Smith* 124

IX. The New Evolutionary Synthesis

- Modeling Modes of Evolution: Comparing Phyletic Gradualism
  and Punctuated Equilibrium, *W.F. McComas and B.J. Alters* 131
- A Computer Simulation of the Evolutionary Rate in Macroevolution
  O.B. Marco and V.S. Lopez 142

X. Glossary 153

XI. References 157

XII. Acknowledgments 165

About the Editor 167

I. INTRODUCTION

There is little argument that the topic of organic evolution is one of the dominant unifying themes in biology, and most agree with geneticist Theodosius Dobzhansky (1973) that "nothing in biology makes sense except in the light of evolution." The place of evolutionary theory within biological history is of such vast importance that it demands a major role in biology education. Knowledge that change through time has occurred and that the proposed mechanisms for such change have withstood the test of scientific scrutiny is vital to students who hope to understand almost any aspect of the life sciences in proper context.

In spite of the strong justification for including evolution-related instruction in biology curricula, "descent with modification" is a particularly difficult educational issue, for by its very nature, evolution is an abstract and generally nonobservable phenomenon. As Stebbins and Allen (1975) stated, "Like the concept of the atom, evolution by natural selection is an abstract principle. It often involves great reaches of time and processes dimly perceptible in ordinary sensation and experience" (p. 206). In addition, many students have only marginally formed the mental structures necessary to conceptualize the complex topics associated with evolutionary biology.

Educators are, therefore, advised to engage learners of evolution theory at the most effective level available — that of the concrete, hands-on laboratory experience. Students will not discover for themselves all of the essential ingredients for the Darwin-Wallace model of evolution by natural selection, nor will they see real evolution in the laboratory. Nevertheless, the wide variety of unique laboratory activities provided in this monograph should make a hands-on approach to evolution education both practical and dynamic, affording classroom instructors the opportunity to put aside the traditional lecture format in favor of a more student-centered investigative approach.

EVOLUTION EDUCATION IN THE LABORATORY

William F. McComas
University of Southern California, Los Angeles, California

The theory of evolution by natural selection was certainly the most important single scientific innovation in the nineteenth century. When all the foolish wind and wit that it raised had blown away, the living world was different because it was seen to be a world in movement. (Jacob Brownowski, 1973)

As a guide in developing this monograph, 58 separate evolution exercises from 12 different current high school biology textbooks were reviewed. Not surprisingly, the number of activities included in laboratory manuals accompanying textbooks corresponds closely with the extent and coverage of evolution in the associated text, and the sophistication and complexity of the laboratory is directly related to the perceived cognitive level and abilities of the target audience. This, of course, makes some laboratory manuals much more useful as sources of evolution-related activities than others. However, few teachers have the luxury of reviewing the dozens of sources currently available and choosing one of the few that features evolution prominently and appropriately.

What is most interesting is the high degree of uniformity in the activities provided in these published sources. For instance, virtually all of the laboratory manuals reviewed include some version of a natural selection simulation, and over half of all the sources consulted include human ancestry activities in which students measure and compare line drawings of various primate skulls. Unfortunately, there is nothing particularly illustrative about the primate skull approach, nor can this approach legitimately be called an inquiry activity. The lesson is clear that most teachers are well advised to look beyond any single source for their evolution laboratory ideas.

One of the most detailed and thought-provoking collections of laboratory activities specific to evolution was produced by the Biological Sciences Curriculum Study (BSCS) as part of its now-classic *Biology Laboratory Block Series* (Webb 1968). Although this publication is no longer in print, BSCS has again performed a significant service to biology education with the development of a new evolution education curriculum module (linked to videodisc images) called *Evolution: Inquiries into Biology and Earth Science* (BSCS 1992).

To help bridge the gap between what is already available in curriculum projects and in classroom laboratory manuals, we endeavored to locate nontextbook, high-quality, inquiry-based activities related to aspects of evolution biology. The activities included in this monograph represent a collection of unique, clever and effective means of addressing evolution in the school science laboratory and come from a number of journals, including *The American Biology Teacher,* the *Journal of Biological Education, School Science Review,* and *Systematic Zoology.* They represent excellent ways for teachers to broaden the range of experiences in which evolution can be made to come alive in the science laboratory.

How to Use this Monograph

Some of the activities in this monograph are previously unpublished exercises; some are new versions of well-known labs; a few would make useful classroom demonstrations; and several require somewhat sophisticated equipment. As a group, these activities represent a valuable teaching resource. Biology educators can now illustrate most aspects of the Darwin-Wallace model of evolution by natural selection by choosing an appropriate activity from each section.

For instance, there are several different exercises requiring students to propose phylogenies of evolutionary trees; each with its own strengths and limitations. The main justification for providing several activities targeting the same evolutionary concept is to provide opportunities to address specific concepts to be experienced by students while maintaining an appropriate level of complexity. With access to several similar activities, educators may use one exercise for instruction and another for authentic assessment. For example, in the laboratory activity by Burns (see p. 120), students use nuts and bolts to propose an evolutionary lineage. Later, another phylogeny exercise by Hageman (see p. 111) using the newly-discovered creatures called "alumenontos" (aluminum pull tabs) could be employed in a test situation to see if students truly can apply what they have learned about evolutionary relationships.

The laboratory activities are edited into a common format and placed together with other exercises illustrating the same evolutionary principle. These major principles include evidences of evolution, variation within the species, adaptation, and simulations of natural selection. Each activity begins with a brief overview so that teachers can judge the suitability of the exercise along with several of the evolutionary principles exemplified by the specific laboratory.

The most appropriate level (life science, general biology, and/or advanced biology) is associated with each activity. These levels relate to the activity as written, but simple modifications might extend the usefulness to students of other abilities. For instance, the addition of the mathematical calculations or the substitution of the Hardy-Weinberg equation for simple ratios might make several of the simulations applicable to life science students. Conversely, the addition of more sophisticated analyses of results could extend many of the activities to the advanced biology classroom.

The materials section is divided between those materials necessary for the entire class (possibly 24-32 students) and those needed for each individual laboratory group within that class. Each student group may consist of any reasonable number of students (i.e., four to six students per group).

Evolution by Natural Selection: A Review

The factors involved in natural selection and the results of the selection process are adapted from those summarized by Huxley (1966) as follows:

- All organisms show considerable natural variation within each species.

- Much of this variation is inherited.

- In nature, all organisms produce more offspring than can survive.

- Accordingly, there is a "struggle for existence" — not all the offspring will be able to survive to reproduce.

- Some variants have a better chance of surviving or reproducing than others.

- The result of the above is natural selection — the differential survival or reproduction of favored variants; and this, given sufficient time, can gradually transform species and produce both detailed adaptation in single species and the large-scale, long-term improvement of types.

This monograph is based on the realization that, while it may not be possible to "see" evolution in the laboratory, it is possible to investigate many of the essential ingredients of the Darwin-Wallace model of evolution by natural selection through appropriate investigative activities.

A Philosophy of Laboratory Instruction

Although the laboratory activities presented here are provided in a traditional format, it is not our intent that each exercise be photocopied in its entirety and simply handed to students. In keeping with the new emphasis on constructivist learning, we recommend that students be given minimal written instruction and challenged instead to investigate the problem with methods of their own design.

Of course, if some new technique needs to be taught in order for students to effectively complete the investigation, teacher-provided instruction is the best choice. However, even new techniques ought to be experienced by students in context rather than followed blindly in a step-by-step fashion.

This is also true of data tables. Although sample tables have been included for your review, generally we advise that students develop their own means for data reporting. Models that make sense to students personally will be much more useful to them than any teacher-designed data reporting sheets. Since, however, this may not be a simple matter for students who have grown accustomed to the more prescriptive types of laboratory exercises, such an instructional technique should be phased in over time.

Basically, what we propose is a learning cycle approach to the laboratory experience, whereby students are provided an opportunity to investigate a problem in the lab before it is discussed in detail in class. (It will be much more interesting for students to discover for themselves that there is variation within species, rather than simply verifying that fact after the concept is presented during a lecture.) After a thorough classroom discussion of their lab results and conclusions, students are asked to explore the phenomenon further by making a prediction or by applying what has been learned. The variety of laboratory exercises targeting the same aspect of evolution theory provided here makes this approach possible.

Acknowledgments

I am grateful to the authors who originally developed the wonderful activities included in this monograph. In addition, I would also like to thank the various publishers who have permitted these exercises to have a "second life" here.

I alone take full responsibility for the editorial decisions that may have resulted in changes in meaning or style when the activities were modified, abridged and extracted from more lengthy articles, and molded into the final common format. To my European colleagues, I have made what I hope is the pardonable sin of modifying British English into the colonial form.

Although the original authors were not given the opportunity to review the changes, I am hopeful that the final monograph will meet with overall approval from those whose ideas are reprinted in its pages. A complete reference is provided with each activity, and readers are encouraged to consult the original when quoting any of these authors.

I appreciate the assistance of the National Association of Biology Teachers and its Publications Committee members in the production of this monograph. I am particularly grateful to Committee member Suzanne Black and my

University of Southern California colleague, Brian Alters, who both reviewed the manuscript and made countless useful suggestions for improvement.

I would also like to thank Brenda Farfán and Cassandra Davis of USC, who did much of the typing, and Sherry Grimm and Michele Bedsaul of the NABT Publications Department for their editorial expertise and untiring support.

Finally, special thanks go to colleagues Alan Colburn and Michael Clough, who wrote the introductory sections.

MISCONCEPTIONS IN EVOLUTION EDUCATION

Alan I. Colburn
University of Iowa, Iowa City, Iowa

How would your students answer the following?

- *Cheetahs can run faster than 60 miles per hour when chasing prey. How would a biologist explain how this ability evolved in cheetahs, assuming their ancestors could only run 20 miles per hour?*

- *Cave salamanders are blind (their eyes are nonfunctional). How would a biologist explain how this inability evolved from sighted ancestors?*

Responses to questions like these, from Bishop and Anderson (1986), provide clues into student views of evolution and natural selection. The responses might be surprising. They suggest the debate between Lamarckism and natural selection is not over. In fact, Lamarckism — the idea that changes acquired during one's lifetime are imprinted on the genes and become a part of the heritage of future generations — may even be more accepted than Darwinian thinking.

Lamarck's idea that generations of giraffes stretching their necks to get food created their present anatomy seems more reasonable to many students than the idea of natural selection. Full appreciation of research about student understandings of ideas like evolution, however, first requires a brief examination of the theory of constructivism.

Constructivism and Student Conceptions of Evolutionary Biology

Science educators are rapidly coming to accept the tenets of a learning theory called "constructivism." Constructivist thinking rests on the assumption that learning is an active process in which the learner constructs ideas to account for new phenomena. To learn something new, a student must literally change his or her mind. And we all know how hard it is to get other people to change their minds!

Constructing ideas about the physical world begins long before school. Learning about the world, of course, neither begins in nor is limited to what happens in classrooms. As a result, students have all sorts of firmly held ideas, concepts and theories about how the world works before they enter our classrooms. These concepts frequently differ from those accepted by scientists.

When the science ideas and vocabulary presented in classrooms conflict with students' intuitive ideas — backed up by years of experience — formal science is usually the loser. Students' present understanding is rather resistant to change, although students can hold two conceptions simultaneously: one just for science class and one for everywhere else. In addition, when instruction does have an effect, the ways that students' views change may be other than what was intended by the teacher.

An important implication of constructivism is that, since students have had different experiences before coming to class, they may interpret instruction differently. In other words, different students may take different things away from the same lesson because of the notions they already had in their minds.

With this information in mind, it is time to address the kinds of conceptions that students hold about the ideas of evolution, natural selection and adaptation.

Student Conceptions of Evolution, Natural Selection and Adaptation

Amazing similarities exist between the ideas students of various ages have about ideas such as evolution, natural selection and adaptation. From children (Renner et al. 1981; Minitzer & Arnaudin 1984) to college students (Bishop & Anderson 1986; Bishop & Anderson 1990) and graduate students (Brumby 1984), people display the same kind of thinking about topics associated with evolutionary biology.

While many students accept that organisms change with time through evolution, students often hold ideas about the mechanisms accounting for the changes that are quite different from those of scientists. Biologists posit that new traits come from seemingly random changes in genetic material, which then survive or disappear due to selection by the environment. The genetic changes (mutation or recombination) occur separately from selection; this is the key point — evolution's mechanism involves two separate, distinct and independent processes.

Many students, on the other hand, think about evolution in Lamarckian terms — a single process affects the development of traits in a population. The environment literally causes traits to change over time. This makes sense on the surface. After all, as environments change, organisms change. People often falsely assume

this correlation implies a cause and effect relationship.

A major reason organisms develop specific traits, in students' minds, is because the organisms need the traits to survive. As a response to Bishop and Anderson's questions, for example, one student wrote that cheetahs needed to run fast for food, so nature allowed them to develop faster running skills. Similarly, organisms developed fur because they needed the warmth to adapt to colder temperatures.

Students also believe organisms change in response to use or disuse of organs or abilities. Species change because members do or do not use these organs and abilities. Thus, cave salamanders' eyes are nonfunctional simply because they do not use them. If this were true, the more we use cars as transportation (rather than walking), the smaller and weaker our legs should become.

Another reason that the researchers conclude students believe the environment causes evolution comes from the various ways people use the terms "adapt" and "adaptation." The words have different meanings in and out of biology. Webster's *New Collegiate Dictionary* defines adapt as "to make fit (as for a specific or new use or situation) often by modification." The definition of adaptation includes the following entry: "Adjustment to environmental conditions: as ... modification of an organism or its parts that makes it more fit for existence under the conditions of its environment."

"Adapt" and "adaptation" have Lamarckian connotations everywhere except in the science classroom. It is no wonder students hear teachers and textbooks talking about adaptation and think in terms of the environment directly affecting organisms. We have indeed adapted, in the everyday sense of the word, to our surroundings, based on our needs. This, however, is cultural, not biological, evolution.

All the ways we have evolved during the last 50,000 years are cultural. The key point to be made here is that cultural evolution generally is Lamarckian, with change occurring through use and disuse. Our cultural evolution has been directed change. And it is this kind of "evolution" knowledge and experience that forms the basis for student conceptions of evolution, selection and adaptation.

Changing Student Thinking

Since students already have many diverse ideas about evolutionary biology, from a constructivist perspective, it will be necessary for teachers to work to change students' minds. Saunders (1992) discusses four interrelated teaching strategies to help students change their minds about topics like evolution.

First, there are hands-on laboratory activities — but not just any activities. The student must test what he/she already knows to form an expectation about what will probably be observed. This type of lab experience is often called "investigative," "inquiry," or "open-ended." Results differing from those expected create disequilibrium in the student's mind. This is the first step for students as they begin questioning their thinking about a topic.

Second, students need active cognitive involvement. This means students have to use their heads. As elementary students, we were all told to put our "thinking caps" on, and students still need to have their thinking caps on to promote cognitive change.

Strategies include having students and teachers thinking aloud, developing alternative explanations, interpreting data, constructively arguing about the phenomena under study, developing alternative hypotheses, designing further experiments to test alternative hypotheses, and choosing hypotheses from competing explanations (Saunders 1992). Students must do the work of

learning; the teacher cannot do it for them. Wisdom simply cannot be taught.

Many of the strategies mentioned above are facilitated through the use of group work, a third way to help students learn from a constructivist perspective. Small group work stimulates thinking, especially if students are explaining or defending their thinking to their peers.

Finally, there is assessment. If teachers do not assess the kind of higher-level thinking they are encouraging when using the strategies above, then students probably will not pay the kind of attention needed for meaningful learning to happen. One way to do this is by asking the kinds of questions that began this article.

Keown (1988) looks at teaching evolution specifically, using a Piagetian framework. He shows how abstract the concepts of evolution are — pointing out, for example, that many students find no less plausible the idea of creating Eve from Adam's rib versus a protozoan changing into an elephant, regardless of the time frame involved. He suggests ways to make concrete background information students need to understand evolution — the geological time scale, continuously changing environments, genetic variation, and the biological potential of organisms to produce virtually unlimited numbers of offspring if left unchecked.

One suggestion in teaching about evolution and natural selection is to be rather careful using the term "adaptation," since students may interpret the word differently than scientists do. One possibility is to replace use of the term with phrases like "inherited changes that help the organism." In other words, simply avoid using the term initially. Alternatively, perhaps it is sufficient to simply refer to biological adaptation, stressing how biological adaptation differs from "everyday" adaptation.

Another suggestion comes from the fact that the

proponents of constructivism place importance on starting with students' ideas. In this case, that would include showing students you understand the appeal of the Lamarckian thinking they may hold. After all, "inheritance of acquired characters" (the phrase biologists often use) does make sense; it is simple; it is gratifying — it seems to place us at an evolutionary pinnacle — and it fits with what students already believe about adaptation. In fact, evolution would proceed more efficiently if it worked the way many students believed! The only problem is that it does not.

Teachers and students both can see the appeal in Lamarckian thinking, but getting students to change their minds requires that they see the flaws in their personal theories and know that accepted scientific wisdom offers a better explanation than their thoughts on how organisms adapt in response to new environments. Perhaps an examination of what changed the minds of the scientific community about this topic is worthwhile since no educational research is yet available to assist teachers.

Refuting the Idea of Inheriting Acquired Characteristics

The idea of inheritance of acquired characteristics was refuted by experimentation and theoretical argument (Mayr 1982). According to Mayr, most of the experiments fell into one of three categories: experiments in the total disuse of a structure, in the amputation of a body part, and in selective breeding among pure lines. In the first two cases, offspring would be expected to have smaller versions of affected structures. In the latter case, offspring would be expected to be different from their parents.

Although students could theoretically replicate experiments involving the effects on offspring of amputating parts of a plant, doing this kind of laboratory work may prove difficult in many classrooms. Instead, however, teachers can challenge students to come up with (and explain) instances from the student's personal experiences that seem to refute the theory. For example, many of us know of deaf people who produced hearing children (not to mention hearing parents who had a deaf child) and people with birth disorders producing unaffected children.

An important part of the theoretical argument against this kind of inheritance came with acceptance of the idea that germ cells are separate from body (somatic) cells. Changes in body cells do not affect germ cells. If students understand this point, they can be challenged to explain how changes in a mature organism can affect its sperm or egg cells.

Another part of the theoretical argument against acquired inheritance is showing that the kinds of phenomena explained by the theory can be explained equally well or better on the basis of Darwinian theory. This is where the teacher's knowledge may play a large part in changing student thinking. You will be the one who helps convince students that Darwinian thinking effectively explains how organisms change through time. You will have to apply your knowledge of natural selection to help students understand the adequacy of this idea (although students who understand the theory can perform similar functions working in small groups with other students).

Constructivism helps explain the difficulty inherent in teaching students complex ideas like those of evolution and natural selection. The theory also offers a powerful framework to view and understand your students. Enhanced understanding of student thinking will lead directly to the most appropriate instructional strategies.

REDUCING RESISTANCE TO EVOLUTION EDUCATION

Michael P. Clough
Memorial High School, Eau Claire, Wisconsin

The citizens' appalling ignorance of the nature of science ... bodes ill for the future. And the more I think about this problem, the more I feel that the fault is mainly ours — we, the teachers in the schools, colleges, and universities of the nation, must accept much of the blame. (John A. Moore, 1983)

Students of all ages not only possess a host of misconceptions concerning biological evolution, but many are secretively or openly hostile toward the topic when it is addressed in science classes. These misconceptions and apprehensions exacerbate the challenges in teaching biological evolution. Therefore, before initiating activities designed to illuminate aspects of biological evolution, such as those provided in this monograph, teachers should first seriously consider the "conceptual baggage" that students bring to this topic.

In the previous section, Colburn addressed some of the biological misconceptions related to evolution education. In this chapter, I will focus on philosophical issues that may block learning, as failing to address all misconceptions will seriously compromise the desired outcomes. What teachers do with activities is at least as important as the activities themselves! This paper suggests strategies that facilitate a more accurate portrayal of the nature of science and diminish hostility toward evolution education, thereby promoting a deeper understanding of evolutionary theory.

The Public Evolution/Creation Controversy and the Nature of Science

Researchers (Carey & Strauss 1970; Rowe 1976; Hodson 1988; Eve & Dunn 1990) have shown that science teachers continue to hold fundamental misconceptions regarding the nature of science. Not surprising, therefore, are the numerous studies documenting science students' misconceptions concerning the nature of science (Horner & Smith 1981; Rowell & Cawthron 1982; Johnson & Peeples 1987; Rubba, Ryan & Aikenhead 1992). John Moore (1983) claims that the public evolution education controversy is, in large part, a result of misunderstandings concerning the nature of science:

> "... It becomes evermore important to understand what is science and what is not. Somehow we have failed to let our students in on that secret. We find, as a consequence, that we have a large and effective group of creationists who seek to scuttle the basic concept of the science of biology ... a huge majority of citizens who, in *fairness*, opt for presenting as equals the *science* of creation and the science of evolutionary biology ... It is hard to think of a more terrible indictment of the way we have taught science."

Johnson and Peeples (1987) found that, as students' understanding of the nature of science increased, they were more likely to accept evolutionary theory. Especially disturbing, then, are the results from the same investigation that show that biology majors have a low understanding of the nature of science. In a smaller study, Scharmann and Harris (1992) found that promoting science teachers' applied understanding of the nature of science reduced anxiety toward the teaching of evolution.

Not surprisingly, one widely accepted component of scientific literacy that has emerged is the need for individuals to have a thorough understanding of the nature of science (ASE 1981; NSTA 1982; AAAS 1989; Matthews 1989; NAEP 1989). The authors of the reports cited above have prompted many science educators to call for increased emphasis on the social studies of science in preservice and inservice science teacher education programs (Nunan 1977; Manuel 1981; Summers 1982; Gallagher 1984; Clough 1989; Matthews 1989).

Suggestions for Reducing Resistance to Evolution Education

No single strategy will pacify all those who oppose evolution education, but a large middle ground of students and parents exists that, while not having strong convictions for any one position, is sympathetic to the "fairness" issue and seriously believes a controversy exists in the scientific community concerning biological evolution. The following suggestions are intended to help science teachers reduce resistance to evolution education, avoid unnecessary controversy, and promote an understanding of the nature of science and biological evolution.

A) Clarify the Distinction Between Biological Evolution and the Origin of Life

Much of the resistance to evolutionary theory arises from the mistaken notion that biological evolution and ideas concerning the origin of life are one in the same. This misconception is held by creationists, the general public and, tragically, by Supreme Court Justices, as evidenced by Justice Scalia's opinion in the Louisiana evolution/creation case. How life arose is an extremely interesting scientific problem. However, biological evolution per se does not involve the study of origins.

"Evolution studies the pathways and mechanisms of organic change following the origin of life" (Gould 1987). This single demarcation often eliminates most resistance to biological evolution. Of course, discussions of cosmology should not be avoided, but teachers should make it clear that biological evolution explains the diversity and similarity of life on this planet — not how life first arose.

B) Use the Language of Science Correctly and Consistently

Science teachers must be very careful with significant language related to the nature of science. Words such as "prove," "true," "theory," "law" and "hypothesis" have different meanings in and out of science. If used incorrectly, these words have the potential to create misconceptions.

For example, students often see scientific ideas as copies of reality (Ryan and Aikenhead 1992). Science teachers create needless trouble when they perpetuate this misconception. Many arguments can be made against the notion of "truth" or "certainty" in science, but Einstein and Infeld (1938, p. 31) provide an easily understood analogy:

"In our endeavor to understand reality, we are somewhat like a man trying to understand the mechanism of a closed watch. If he is ingenious, he may form some picture of a mechanism which could be responsible for all the things he observes, but he may never be quite sure his picture is the only one which could explain his observations. He will never be able to compare his picture with the real mechanism, and he cannot even imagine the possibility or the meaning of such a comparison."

Because the "watch" can never be opened, asking whether ideas concerning the natural world are true (i.e., copies of reality) is to ask an unanswerable question. Einstein suggested a different view of scientific truth — truth is what works! Science teachers should acknowledge that all scientific ideas (not just evolution) are tentative, are open to revision, and are judged by how well they work.

As a second example, consider that outside of science the word "theory" most often means "guess" or "speculation." Ryan and Aikenhead (1992), after collecting the responses of more than 2,000 grade 11 and 12 students, found:

"The majority [of students] (64%) expressed a simplistic hierarchical relationship in which hypotheses become theories, and theories become laws, depending on the amount of *proof behind the idea*."

When individuals bring this misconception to the evolution/creation controversy, nonsensical statements such as "evolution is *only* a theory" are often heard. The word "theory," however, has an entirely different meaning in science. Among other things, scientific theories predict, explain (Campbell 1953), and provide conceptual frameworks for further research (Kuhn 1970). Certainly some scientific theories are more speculative than others, but all must perform the functions just described.

Finally, due to the emotional response of many students toward evolutionary theory, what science teachers say and how they say it are especially critical. A fundamental tenet of constructivist learning theory is that students' views, whether they be alternate conceptions or misconceptions, must be treated with great respect. Making light of students' views only exacerbates the difficulty of persuading them to build a functional understanding and acceptance of biological evolution.

C) Stress Functional Understanding Rather Than Belief

When students are faced with a choice between evolutionary theory and their personal religious conviction, science will most certainly lose. Lawson and Worsnop (1992) write:

"… Every teacher who has addressed the issue of special creation and evolution in the classroom already knows that highly religious students are not likely to change their belief in special creation as a consequence of relatively brief lessons on evolution. Our suggestion is that it is best not to try to do so, not directly at least."

Students are more likely to consider and accept evolution if a functional understanding of the theory is stressed. This, once again, can be accomplished by showing how the theory works (i.e., predicts, explains and provides a framework to conduct further research).

A recent book, *Science as a Way of Knowing*, by John Moore (1993) does just this by providing a comprehensive list of deductions that follow from evolutionary theory and the evidence sustaining these deductions. These deductions represent propositions derived from and supported by evolution by natural selection. Students might be challenged to develop their own deductions and then investigate to see if evidence supports those deductions. According to Moore (1993), there are a number of deductions, including the ideas that:

- The species that lived in the remote past must be different from the species today.

- The older the sedimentary strata, the less the chance of finding fossils of contemporary species.

- The simplest organisms would be found in the

very oldest fossiliferous strata, and the more complex ones only in more recent strata.

- It must be possible to demonstrate the slow change of one species into another.

- It must be possible to demonstrate forms between major groups (e.g., phyla, classes, orders) should have existed.

- The age of the Earth must be very great.

- If the members of a taxonomic unit share common ancestry, it should be reflected in their structure and embryonic development.

- If a unity of life is the basis of descent from a common ancestry, then this should be reflected in the structure of cells and in the molecular processes of organisms.

Another particularly effective approach to show the usefulness of evolutionary theory is to examine the implications of evolutionary theory for modern medicine. For example, an increasing number of biotechnology companies are using "applied" molecular evolution in the development of high-tech drugs (Bishop 1992). At a macroevolution level, Gould (1988) writes a particularly biting attack, showing how an ignorance of evolutionary theory resulted in a questionable heart transplant from a baboon to a human infant.

Some individuals will argue that the instrumentalist perspective I am advocating is simply a form of avoidance. However, if we look to the history of science, well-known scientists took this same approach. In 1867, Kekule reputedly wrote: "I have no hesitation in saying that, from a philosophical point of view, I do not believe in the actual existence of atoms … As a chemist, however, I regard the assumption of atoms as absolutely necessary."

Students, when presented an instrumentalist

perspective, not only are more likely to consider evolution but also will learn a great deal about the history, philosophy and sociology of science.

D) Realize Knowledge Is Not Democratic

The public clamor for teaching both evolution and creationism in our nation's public science classrooms is, in part, an admirable, but sometimes inappropriate, belief in fair play. However, fair play doesn't mean giving credibility to every idea. We don't allow discredited views, such as a flat-earth, astrology, Aristotelian physics, and geocentricity, into our science curriculum simply because a significant number of citizens may believe these ideas. Students need to be made aware that the scientific community, not public opinion polls, decides what is good science!

E) Realize Science Provides Natural Explanations for Phenomena

Ryan and Aikenhead's (1992) research indicates that 46% of students hold the view that "science could rest on the assumption of an interfering deity." This misconception concerning a basic assumption of science has devastating consequences as students interpret the meaning of data gathered in evolution activities. This view may be confronted in the following way:

> A popular science cartoon by Sidney Harris (1977) has two scientists at a blackboard considering a lengthy series of mathematical computations interrupted by the written statement, "Then a miracle occurs," followed by another series of computations. The scientist in the foreground, pointing at the reference to the miracle, states, "I think you should be more explicit here in Step 2."

This cartoon conveys an important message about science as well as the evolution/creation controversy. Science deals with the natural world and, consequently, its explanations must be couched in natural expressions. Explanations employing supernatural events and/or deities are beyond nature and, hence, beyond the realm of science.

What the above comic points to is the crucial concept that references to the supernatural are not particularly useful in science. Creationists do not seem to understand that, even if the evidence did suggest an abandonment of biological evolution, the scientific community would then attempt to explain the diversity of life in other naturalistic terms. This is the essence of science.

F) Remember That "Fitness" Is Not Just "Differential Reproductive Success"

Perhaps the most intuitive aspect of biological evolution is *natural selection* — the idea that organisms with advantageous characteristics in a given environment have a greater chance of survival and reproduction than organisms lacking these characteristics. Herbert Spencer coined the popular phrase "survival of the fittest" as a definition for natural selection.

Many laboratory activities address this fundamental concept. However, when "fitness" is defined solely as "differential reproductive success," the phrase "survival of the fittest" becomes "survival of those who survive" — an empty tautology (Gould 1977). Tautologies (e.g., my mother is a woman) are true by definition and, hence, not open to testing. Alert creationists will, with good reason, attack evolution on the grounds that tautologies are not testable. The solution to this apparent problem is that "fitness," while often expressed as differential survival, is not defined by it (Gould 1977).

G) Consider the Issue of Falsifiability

Creationists often claim that evolutionary theory is not falsifiable and, hence, not science by Popper's (1963) criteria. Yet, in the next breath, they will cite several pieces of evidence that

supposedly falsify evolution. These two positions are self-contradictory because an idea cannot be both unfalsifiable and falsified.

H) Consider Anomalies in Science

Perhaps the most counterintuitive notion that comes from the nature of science is the well-supported view that unsolved puzzles and seemingly refuting evidence do not always result in rejection of a scientific idea. Kemp (1988) writes:

> "Any theory of the scope of the theory of evolution will always be faced with anomalies, things that it cannot explain, or even things that seem to contradict it."

The reasons for this are varied and detailed, but comprehensive theories are not discarded simply because several pieces do not fit. Many historical examples exist demonstrating that contradictory data did not result in abandonment of ideas accepted today as good science (Kuhn 1970; Chalmers 1982; Kitcher 1982).

The debate surrounding punctuated equilibrium is a recent example illustrating that anomalies do not always result in abandonment of well-supported ideas. Some scientists have always thought that the geological record, although replete with transitional fossils, was not as rich as might be expected. This potential anomaly, although not seen as such by all scientists, in no way diminishes the idea that evolution has taken place. Rather, accommodations were made elsewhere to account for this apparent anomaly.

Punctuated equilibrium is in perfect accord with biological evolution, and it accounts for what its proponents believe are an insufficient number of transitional fossils. Understanding the role of anomalies in science is critical for students as they examine the evidence for evolution, and as they perform laboratory investigations.

Summary

Too many students graduate without gaining a sufficient understanding of biological evolution. Evolution must be taught in order to accurately portray modern biology and prepare students for the future. Much of the resistance to evolution education by students, parents and teachers can be attributed to a poor understanding of the nature of science.

I have argued elsewhere (Clough 1989) that science teachers have a profound effect on students' understanding of the nature of science and thus are responsible for:

1. Expressing potential explicit and implicit views of the nature of science portrayed in science activities.

2. Modifying existing activities so that they more adequately portray the nature of science.

3. Evaluating textbooks, audiovisual materials, and other curriculum materials for their accuracy in portraying the nature of science.

4. Demarcating (using consensus views from the social studies of science) science, nonscience and pseudoscience.

5. Implementing correct historical examples (where appropriate) that effectively convey a more accurate portrayal of the nature of science.

The suggestions made here are important because laboratory activities directed towards illuminating biological evolution are only as good as the teachers who implement them. Model activities will be severely compromised without exemplary teaching.

Genuine acceptance of evolutionary theory first requires a functional understanding of the idea of evolution, and this is preceded by an openness to learning about it. Science teachers, by utilizing strategies described here, can increase their students' understanding of the nature of science and significantly improve their attitudes toward evolution education. This will pave the way for full engagement in all instruction devoted to one of the most comprehensive frameworks created by human intellect.

II. EVIDENCES OF EVOLUTION

The idea that evolution has occurred is not intuitively obvious. However, as curious individuals began to explore the natural world, a number of intriguing observations led to the inescapable conclusion that all living things are related to each other. Evolution as an idea has roots that may be traced to ancient Greece, and the evidence that has accumulated since that time in support of evolution is abundant.

Many laboratory manuals include activities in which students compare fossils with living forms or where they study, compare and contrast apparently unrelated organisms (such as starfish and humans) as evidence of evolution. The two activities in this section provide additional opportunities for students to visualize the principle of *homology* — the presence of structures or biochemicals that developed from those in a common ancestor and are now found within two distinct living forms. Homology is frequently cited as support for the idea that evolution has indeed occurred.

In the first activity in this section, students can discover the relationship between plants of different species by examining the nature of their shared biochemicals. In the second exercise, structural homology is demonstrated as students look for commonalities in the anatomy of various vertebrates. In both cases, evolution is the explanation for the existence of common structures and chemicals in otherwise unrelated species.

BIOCHEMICAL EVIDENCE FOR EVOLUTION

Based on an original activity by
Daniel J. Dyman

Thin layer chromatography (TLC) of plant tissue extracts is used to compare the degree of similarity of plants in the same genus. The possibility of using plants with relationships unknown to the students or of exploring student-initiated lines of research are also included.

Evolutionary Principles Illustrated

- Biochemical homology
- Naturalistic taxonomy

Introduction

Thin layer chromatography (TLC) is a technique that can be conveniently used in the laboratory to generate evidence supporting the principle that degrees of biochemical similarity reflect degrees of evolutionary relatedness among organisms. When TLC is applied to the analysis of tissue extracts of various organisms, it can be shown that similarities among the extracts result from an ancestral relationship.

Intended Audience

- General biology
- Advanced biology

Materials (for each student group)

- leaves of several different species of plants in the same genus, such as *Erythronium* (adder's tongue, trout lily or dogtooth violet) or *Trifo-*

lium (white, red, alsike or crimson clover)
(Note: Other genera of plants could be investigated as part of this activity.)
- methanol (extracting agent)
- concentrated HCl [12m] (extracting agent)
- 5x11-cm glass plates or microscope slides
- canning jar large enough to hold the glass plate with lid
- silica gel (14 g)
- methanol-chloroform solution (20 ml)(3:7 v/v)
- microliter syringe or 50-μl microcapillary tube
- needle or dissection probe
- 1 glass rod
- labeling tape

Materials (to be shared by all class members)

- 125-ml Erlenmeyer flask
- drying oven
- ultraviolet lamp (longwave)
- distilled water (40 ml)

Safety Note

Although the required chemicals are found in many school laboratories, several are flammable, and the concentrated HCL is corrosive. Work with supervision in a fume hood or well-ventilated room, with all flames extinguished. Read and understand all safety instructions for the proper handling of these substances.

Procedure

The organisms used in this biochemical investigation of evolution are *Erythronium americanum, E. albinum, Trifolium repens, T. pratense,* and *T. arvense.* Plants of these two genera are used because of their common occurrence and because students can easily relate them morphologically. (It may be necessary to secure these plants from a biological supply company or locate them in the environment with the assistance of a field guide.)

Part I: Preparing the Plant Extracts

Collect the plants, excluding the roots, when they are in flower. Wash the plants, superficially dry them, place them between layers of newspaper, and air dry or oven dry them at 45°C or less.

Obtain plant-tissue extract by placing approximately 0.4 g of each of the air-dried plants in a drying oven at 45°C for approximately 12 hours. Pulverize the oven-dried plant tissues with a glass stirring rod after placing them in small glass vials. Add 2.5 ml of extracting agent, a methanol-concentrated hydrochloric acid solution (99:1 v/v), to each of the vials containing the pulverized plant tissue. Seal the vials and place them in the dark at room temperature for 12 hours. The resulting plant-tissue extract may be stored in a refrigerator for several days.

Part II: Preparing the Thin Layer Chromatography Plates

Prepare the gel by placing 14 g of silica gel H for TLC in a 125-ml Erlenmeyer flask, adding 40 ml of distilled water, and swirling the gel-water mixture. Tape the long edges of scrupulously clean 5x11-cm glass plates with 1.2-cm moisture-resistant labeling tape. The tape serves as a gauge for limiting the thickness of the gel layer to be distributed on the surface of the plate.

Prepare the plates any of the following ways:

• Dip the glass plates into the gel-water mixture.

• Spray the glass plates with gel-water mixture.

• Spread the gel-water mixture over the glass plates. (The third method is suggested for this investigation.)

Quickly pour approximately 2 ml of the mixture onto each of the taped glass plates. Evenly distribute ("strike off") the gel with a 1-cm diameter glass rod. In distributing the silica gel, an even layer is desirable. Carefully remove the tape from the plates, and the allow the gel layer to air-dry.

Regardless of the method used to coat the glass plates with the gel layer, activate the plates by placing them in a drying oven at 95–100° C for 30 minutes. Cool the plates to room temperature before using. The activated plates may be stored for several days if kept in a dry, dust-free environment.

Part III: Applying the Plant Extract

Make a microcapillary tube by gently heating the midregion of a 50-µl pipette. As the pipette is heated, draw the ends manually apart to produce a narrow, constricted region. Then break the pipette in the region of the constriction to produce two microcapillary tubes, each having a tiny internal diameter.

With the use of the microcapillary tube, make a band across the narrow edge of the silica gel. The band should be approximately 1.5 cm from the bottom edge of the plate. Three applications of the extract is ideal. An insufficient concentration of extract results in a separation that is not readily apparent; an excessive concentration results in an ill-defined separation. Label the silica gel plates at the top by scratching into the gel layer with a needle or a dissection probe.

Part IV: Developing the Plates

Develop the extract-spotted silica gel plates in a chromatographic chamber, which may be a

Fig. 1. The chromatographic system is shown with a TLC plate in the process of development. The developing chamber is a canning jar with its lid reversed.

sophisticated commercial variety but could be a 1-pint canning jar. If a canning jar is used, reverse the lid so that the rubber sealing ring is up, since typical TLC solvents tend to dissolve rubber.

The recommended chromatographic solvent is a methanol-chloroform solution. The solvent is added to the TLC developing chamber to a depth of approximately 0.5 cm. Caution the students that the solvent level must not exceed the level of the extract band on the TLC plate. If the solvent level touches the extract band, the extract will dissolve into the solvent, and the TLC plate will be ruined.

Place the extract-spotted plates in the developing chamber containing the solvent, seal the chamber, and develop the plate (Figure 1). Development of a TLC plate takes about 20 minutes. The solvent should be allowed to rise to a height of approximately 9-10 cm.

Use a longwave ultraviolet lamp to visualize the developed TLC plates. (Figures 2 and 3 show

the results that can be expected.) The violet bands that appear on the actual plates represent free amino acids. The amino acids are less significant than the secondary compounds as indicators of evolutionary relatedness.

With regard to both the secondary compounds and the free amino acids, students should examine the TLC plates for similarities and

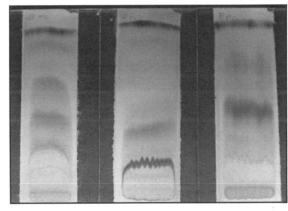

Fig. 2. TLC plates of E. albinum (left) and E. americanum (right) under ultraviolet light. The TLC plates reveal two similar secondary compound bands having Rf values of 0.51 and 0.41, respectively. The photo has been slightly retouched to bring out the chromatographic bands.

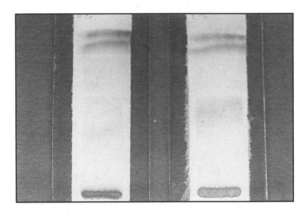

Fig. 3. TLC plates of T. repens (left), T. pratense (center), and T. arvense (right) under ultraviolet light. The TLC plates reveal that T. repens and T. arvense have a common secondary compound band with an Rf value of 0.50 and that T. pratense and T. arvense have a common secondary compound band with an Rf value of 0.26. A common secondary compound apparently does not exist for the three Trifolium species used in this study.

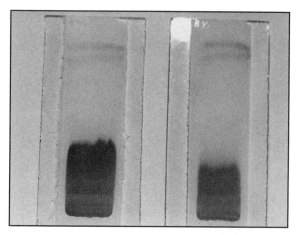

Fig. 4. The nynhydrin-treated TLC plates of E. albinum (left) and E. americanum (right) reveal at least two common amino acid bands, which have Rf values of 0.27 and 0.11.

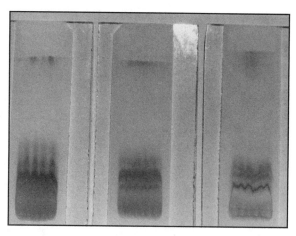

Fig. 5. The ninhydrin-treated TLC plates of T. repens (left), T. pratense (center), and T. arvense (right) reveal at least two common amino acid bands, which have Rf values of 0.18 and 0.29.

differences of various bands. Color and location of the various bands should be taken into account. The degree of ancestral relatedness is reflected by the degree of similarity represented by the separation.

Calculate *ratio–to–front* (Rf) values for each of the bands. The formula used for this calculation is the distance the band traveled divided by the distance the solvent traveled. Rf values are helpful in comparing location similarities that may exist among the various bands. Similar compounds have similar Rf values.

Discussion

The results of this investigation indicate that *Erythronium albinum* and *E. americanum* are more closely related to each other than to any of the *Trifolium* species. Similarly, *T. repens, T. pratense* and *T. arvense* are more closely related

to each other than to either of the *Erythronium* species (Figures 4 and 5).

Add an investigative character to this activity by including an "unknown" plant extract. The unknown can be any of the five plants used in this lab activity or another closely related species the students have not yet investigated. Ask students to associate the unknown with its apparent relative, or the unknown can be an organism that is not morphologically similar to any of the known organisms. Students can note the obvious differences that exist among the various group representatives and the unknown.

Reference

This activity is based on an original exercise by D.J. Dyman (1974). Biochemical lab activity supports evolution theory. *The American Biology Teacher, 36*(6), 357-359, and is modified and reprinted with permission of the publisher.

EVIDENCES OF EVOLUTION THROUGH COMPARATIVE ANATOMY

Based on an original activity by
Robert A. Coler

The skeletal systems of various animals' comparative anatomy are used here to demonstrate evolutionary relationships by providing evidence relating to morphological adaptations in skeletal structures.

Evolutionary Principles Illustrated

• Homology as evidence of evolution
• Comparative anatomy

Introduction

This activity, which may be adjusted to suit student ability, provides an opportunity for students to label and keep track of a particular structure as it is modified by evolutionary forces. The idea that ancestral structures have been reused in descendents is called *homology*. Homology is one of the major pieces of evidence in support of evolution.

Intended Audience

• General biology
• Advanced biology

Materials (for each student group)

• fine-point permanent markers (5 colors)
• tie-on tags (optional)
• clear nail polish
• skeletons of various animals in related classes, such as a frog, lizard, bird, bat and human
• anatomy guides for animals used in the lab

Procedure

To demonstrate evolutionary trends in the laboratory, number the homologous areas on various skeletons with fine-point markers, using a different color for bones, foramina (openings in or passages through bones) and fossae (depressed areas), and tuberosities (elevations or bumps) and processes (prominences or projections). Cover the labels with clear nail polish to prevent smudging. In the case of wet cartilaginous skeletons of animals such as sharks, color–coded cloth laundry tags with piercing clamps may be used. Tie-on tags may be used to avoid permanent labeling if the skeletons are to be used with various student groups.

Next, use the same numbers or names to label homologous areas on the skeletons of related animal groups. If a bone is found with no homology in the other classes, give it a new number or name. The resulting case study will demonstrate the ways in which structures are adapted for new purposes by natural selection.

Figure 1 is an example of this technique, comparing structures in the forelimbs of various organisms. Here, students could be asked to label the phalanges ("finger bones") of each animal and note the changes that have occurred through time.

In the more sophisticated example suggested by Coler (1966), advanced students explore the evolution of the primitive mandibular support

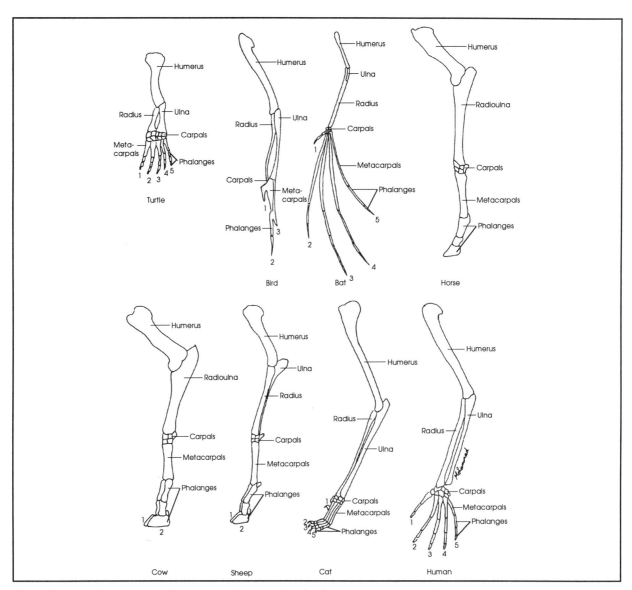

Fig.1. Comparative anatomy: Homologous bones in the forelimbs of several vertebrates.

system into elements of the sound detection organ in more recent animals.

Students might trace the refinement in sensitivity to sound by labeling the *hyomandibular bone* of the *hyoid arch*, which gave rise in the post placoderms to the *urodele eucolumella* and the reptilian *paracolumella*, which was also formed from the *quadrate* and the *mammalian stapes*. Similiarly, the *quadrate* and *articular* in am-

phibians gave rise to the *incus* and *malleus*, respectively, in mammals.

Reference

This exercise is modified from an activity by R.A. Coler (1966). An evolutionary approach to a comparative anatomy laboratory. *The American Biology Teacher, 28*(4), 305-6, and is reprinted with permission of the publisher.

III. GENERAL EVOLUTIONARY PRINCIPLES

In addition to the formal aspect of the Darwin-Wallace model of evolution by natural selection, a number of other issues are central to student understanding of this dynamic process. Two such items are included in this section.

One of the most significant impediments to full comprehension of organic evolution is an appreciation of the immense amount of time that has passed since the origin of life on Earth. During Darwin's day, one of the central objections to the concept of evolution by natural selection was the widespread belief that the Earth was only 6000 years old. Even today, many people still deny the antiquity of the planet and use that issue as a way to reject evolution.

In an attempt to make the issue of geologic time more acceptable, one the activities in this section enables students to construct a scale model of geologic time linked to markers of significant geologic, geochemical and biologic events. Taking a walk through geologic time will enable students to appreciate more fully the vastness of time itself and to make conjectures about relationships between a variety of physical and biological events.

The other activity presented here targets the idea of competitive exclusion — the notion that one species may be more successful in one environment than another and that the successful species will replace the unsuccessful species if the two populations live together. This exercise shows students that fitness operates not only at the level of the individual but also at the level of the entire population. In the case presented here, one entire species is more or less fit than the other.

HOW LONG IS A LONG TIME?
(Constructing a Scale Model of the Development of Life on Earth)

Based on an original activity by
William F. McComas

In this activity, students construct a scale model of geologic time and place markers for significant biologic and geologic events within the model. This exercise will allow students to gain some perspective of the magnitude of geologic time that permits the evolution of the wide variety of life forms that have developed. This model will also provide the opportunity to infer interrelationships between biologic, geologic and chemical events.

Evolutionary Principles Illustrated

- Geologic time
- Major biologic, geologic and chemical events
- Interrelationships between biologic, geologic and chemical events

Introduction

The vast number of years that have passed since the origin of the Earth have permitted a wide variety of events to occur that are of interest to scientists. Students will construct a scale model permitting a leisurely stroll through an enormous expanse of time reduced to the size of a football field.

It is easy to say that the first living cells appeared on Earth about 3.5 billion years ago, but few can really visualize the size of a number as large as 3.5 billion. To further complicate the issue, it is practically impossible to illustrate the expanse of geologic time to scale in a textbook diagram since a division even as small as a centimeter — used to represent one of the geologic periods — would result in a chart many meters long.

For instance, on most textbook geologic time-tables, it appears as if the Pleistocene epoch and the Devonian period lasted the same length of time because they take up the same amount of space. In reality, the Devonian was almost 25 times longer than the Pleistocene. Such is the problem of scale when billions of years are reduced to a single page.

Intended Audience

- Appropriate for all students

Materials (to be shared by all class members)

- geologic timetable
- chart showing significant biologic events
- chart showing significant geologic events
- 65 or more 5x7-inch cards
- metric tape measures (several per class)
- black, green and red marking pens
- cards on wooden stakes on which students may draw or paste pictures of the various significant events

Procedure

1. Decide on the length of the space in which you will set up your scale model of geologic time. The example here is based on a football field that is 91.44 meters long.

2. Calculate the scaling factor for your model by referring to the sample calculation.

3. Use colored markers to label the 5x7 cards. Label the geologic periods with the black marker, the significant biologic events with the green marker, and the geologic events with the red marker.

4. Fold the 5x7 cards as indicated in Figure 1 below to form "tent" shapes. They will then stand up by themselves on the ground.

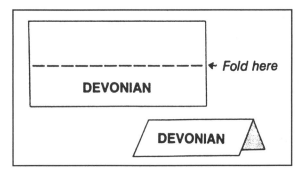

Fig. 1. Example of folded card.

5. Refer to the geologic time scale following this activity (Tables 1a and 1b) and use the scaling factor to calculate the distance for placing markers for the geologic periods.

6. As you calculate positions for the placement of the markers, refer to the charts of the geologic and atmospheric events (Tables 2a and 2b) and the biologic events (Tables 3a, 3b).

7. Use the measuring tape to determine the distance from the goal line for the placement of each card.

8. You are now ready for a walk through an accurate scale model of geologic time.

Discussion

A better understanding of geologic time and the relationship between past events can be gained with the construction of an accurate three–dimensional scale model. In the model discussed here, units of time will be represented by units of distance. Markers placed at certain intervals, corresponding to significant biologic and geologic events, will make it possible to see what events preceded others and how much time or, in this case, distance passed between them. It is also possible for students to infer causal links between biologic and geologic events by visualizing them together.

In this model, the length of a football field is used to represent the length of time that has passed since the formation of the Earth. The calculation below shows that 1 million years on our model will be represented by a distance of 0.01988 m or 1.988 cm. A marker placed on the far goal line of a football field to represent the origin of the Earth would be 91.44 meters away, representing 4600 million (4.6 billion) years.

91.44 meters/4600 million years = 0.01988 meters per million years (m/myrs)

This calculation will convert into meters the number of years that have passed from any past event. For instance, using the knowledge that multicelled plants and animals arose about 700 million years ago, it is possible to determine the proper placement for the marker as follows:

(700 million)(0.01988 m/myrs) = 13.92 meters

The marker for this event should be placed 13.92 meters from the time zero goal line, the one that represents the present. The boundaries of the geologic periods would be determined in exactly the same manner. The Devonian period began about 408 million years ago, or approximately 8.1 meters from the closest goal line.

With some of the more recent events, time intervals are represented in units of less than one meter. This is no problem in the metric system since units can be easily converted from one into another — a feat not so easily accomplished in the English system of measurement. To use the Pliocene epoch as an example, the marker would be placed 0.1 m (10 cms) from the closest goal since the Pliocene began only 5 million years ago.

Once the markers are all in place, it will be possible to "walk" through time from one end of the football field to the other and discuss those events and geologic periods in the order in which they occurred and in the proper scale.

It is possible to adapt this idea to show the even longer period of time since the development of the universe itself, but the scale would have to be recalculated to allow all the events to fit into the football field. This may not be as effective, since the event markers will now be much closer together. This approach is best reserved for a longer area.

Depending on the nature of the group, students could work independently on the calculations or the teacher could provide them. The list of events could be made longer or shorter to suit specific instructional purposes, and the marker cards could be made by the students or prepared in advance by the instructor. With younger or less able students, it may be possible to communicate the point with just a few cards on the scale model.

Students should be encouraged to draw sign-posts for the most interesting events. A picture of a fish representing the Devonian attached to the model or a dinosaur indicating the Cretaceous can make this quite a colorful activity

Two references that are strongly recommended to help fill in the details for both students and teachers are *Life on Earth* by David Attenborough (1979) and *The Rise of Life* by John Reader (1986). These books are both well written and contain beautiful illustrations to tell the story of life's development in an engaging and intelligent way. The chronological treatment used makes both books quite useful for the purposes of extending this activity.

For a visual representation of this approach you might want to show a videotape of part of the episode titled "One Voice in the Cosmic Fugue" from Carl Sagan's *Cosmos* television series.

Reference

The activity presented here was a national award winner in the *Biology Laboratory Exchange Program* sponsored by Prentice Hall Educational Book Division, Englewood Cliffs, NJ.

This activity is based on an activity by W.F. McComas (1988). How long is a long time? *The American Biology Teacher, 52*(3), 161–167, and is modified and reprinted with permission of the publisher.

Table 1a. Student Work Sheet/Geologic Timetable

ERA	PERIOD	EPOCH	BEGINNING (Millions of years ago)	DURATION (Millions of years ago)	NUMBER OF METERS	MAJOR EVENTS
Cenozoic Era	Quaternary	Recent	Began 10,000 years ago			Civilization spreads. Human beings are the dominant form of life
		Pleistocene	2	2		"The Ice Age." Modern human beings present. Mammoths and other such animals become extinct.
	Tertiary	Pliocene	5	3		Fossil evidence of ancient human beings near the end of the epoch. Many birds, mammals, and sea life similar to modern types. Climate cools.
		Miocene	24	19		Many grazing animals. Flowering plants and trees similar to modern types.
		Oligocene	37	13		Fossil evidence of primitive apes. Elephants, camels, and horses develop. Climate generally mild.
		Eocene	58	21		Fossil evidence of a small horse. Grasslands and forests present. Many small mammals and larger mammals, such as primitive whales, rhinoceroses, and monkeys.
		Paleocene	67	9		Flowering plants and small mammals abundant. Many different climates exist.
Mesozoic Era	Cretaceous		144	77		First fossil evidence of flowering plants and trees. Many small mammals. Dinosaurs are extinct by the end of the period. Coal swamps develop.
	Jurassic		208	64		First fossil evidence of feathered birds and mammals. Many dinosaurs roam the Earth.
	Triassic		245	37		Beginning of the "Age of Dinosaurs." Insects plentiful. Cone-bearing plants present.
Paleozoic Era	Permian		256	41		First evidence of seed plants. Fish, amphibians and giant insects present.
	Carboniferous	Pennsylvanian Period	330	44		First evidence of reptiles. Many amphibians and giant insects present. Many large fern trees. Swamps cover many lowland areas.
		Mississippian Period	360	30		
	Devonian		408	48		"Age of Fish." First fossil evidence of amphibians and insects. Many different kinds of fish in the Earth's waters. The first forests grow in swamps.
	Silurian		438	30		First evidence of land plants. Algae, trilobites, and armored fish plentiful. Coral reefs form.
	Ordovician		505	67		Fossil evidence of jawless fish. Algae and trilobites plentiful. Great floods fover most of North America.
	Cambrian		540	35		"Age of Invertebrates." Fossil evidence of trilobites, clams, snails, and seaweed. Seas spread across North America.
	Precambrian	Proterozoic Era	4.6 billion	Almost 4 billion		Fossil evidence of bacteria and algae. Earth forms.
		Archeozoic Era				

Table 1b. Student Work Sheet/Geologic Timetable — Answer Key

ERA	PERIOD	EPOCH	BEGINNING (Millions of years ago)	DURATION (Millions of years ago)	NUMBER OF METERS	MAJOR EVENTS
Cenozoic Era	Quaternary	Recent	Began 10,000 years ago		0.0002	Civilization spreads. Human beings are the dominant form of life
Cenozoic Era	Quaternary	Pleistocene	2	2	0.04	"The Ice Age." Modern human beings present. Mammoths and other such animals become extinct.
Cenozoic Era	Tertiary	Pliocene	5	3	0.10	Fossil evidence of ancient human beings near the end of the epoch. Many birds, mammals, and sea life similar to modern types. Climate cools.
Cenozoic Era	Tertiary	Miocene	24	19	0.4	Many grazing animals. Flowering plants and trees similar to modern types.
Cenozoic Era	Tertiary	Oligocene	37	13	0.72	Fossil evidence of primitive apes. Elephants, camels, and horses develop. Climate generally mild.
Cenozoic Era	Tertiary	Eocene	58	21	1.15	Fossil evidence of a small horse. Grasslands and forests present. Many small mammals and larger mammals, such as primitive whales, rhinoceroses, and monkeys.
Cenozoic Era	Tertiary	Paleocene	67	9	1.33	Flowering plants and small mammals abundant. Many different climates exist.
Mesozoic Era	Cretaceous		144	77	2.86	First fossil evidence of flowering plants and trees. Many small mammals. Dinosaurs are extinct by the end of the period. Coal swamps develop.
Mesozoic Era	Jurassic		208	64	4.14	First fossil evidence of feathered birds and mammals. Many dinosaurs roam the Earth.
Mesozoic Era	Triassic		245	37	4.87	Beginning of the "Age of Dinosaurs." Insects plentiful. Cone-bearing plants present.
Paleozoic Era	Permian		256	41	5.69	First evidence of seed plants. Fish, amphibians and giant insects present.
Paleozoic Era	Carboniferous	Pennsylvanian Period	330	44	6.56	First evidence of reptiles. Many amphibians and giant insects present. Many large fern trees. Swamps cover many lowland areas.
Paleozoic Era	Carboniferous	Mississippian Period	360	30	7.16	
Paleozoic Era	Devonian		408	48	8.11	"Age of Fish." First fossil evidence of amphibians and insects. Many different kinds of fish in the Earth's waters. The first forests grow in swamps.
Paleozoic Era	Silurian		438	30	8.71	First evidence of land plants. Algae, trilobites, and armored fish plentiful. Coral reefs form.
Paleozoic Era	Ordovician		505	67	10.04	Fossil evidence of jawless fish. Algae and trilobites plentiful. Great floods fover most of North America.
Paleozoic Era	Cambrian		540	35	10.74	"Age of Invertebrates." Fossil evidence of trilobites, clams, snails, and seaweed. Seas spread across North America.
	Precambrian	Proterozoic Era	4.6 billion	Almost 4 billion	91.45	Fossil evidence of bacteria and algae. Earth forms.
	Precambrian	Archeozoic Era				

Table 2a. Student Work Sheet/Geologic and Atmospheric Events

Event	Millions of Years Ago	Number of Meters from Closest Goal
Worldwide Glaciations (average)	1.6	
Linking of North and South America with land bridge	5.7	
Formation of the Himalaya Mountains	15	
Collision of Indian and Asian Plates	35	
Separation of Australia and Antarctica	50	
Formation of the Alps	65	
Formation of the Rocky Mountains	70	
Opening of the Atlantic Ocean as the Eastern Hemisphere splits from the West	100	
Formation of Supercontinent – Pangea II	200	
Formation of coal deposits	340	
Oxygen reaches 20% (present level)	380	
Development of the Applachian Mountains	575	
Breakup of the Early Supercontinent	580	
Free oxygen reaches 2% in the atmosphere	600	
Formation of the Early Supercontinent	1250	
Free oxygen begins to build up	2500	
Period of no free oxygen	3700	
Oldest Earth rocks	3800	
Origin of the Earth as a solid mass	4600	

Event	Millions of Years Ago	Number of Meters from Closest Goal
Worldwide Glaciations (average)	1.6	0.032
Linking of North and South America with land bridge	5.7	0.113
Formation of the Himalaya Mountains	15	0.30
Collision of Indian and Asian Plates	35	0.70
Separation of Australia and Antarctica	50	0.99
Formation of the Alps	65	1.29
Formation of the Rocky Mountains	70	1.39
Opening of the Atlantic Ocean as the Eastern Hemisphere splits from the West	100	1.99
Formation of Supercontinent – Pangea II	200	3.98
Formation of coal deposits	340	6.76
Oxygen reaches 20% (present level)	380	7.55
Development of the Applachian Mountains	575	11.43
Breakup of the Early Supercontinent	580	11.53
Free oxygen reaches 2% in the atmosphere	600	11.93
Formation of the Early Supercontinent	1250	42.85
Free oxygen begins to build up	2500	49.70
Period of no free oxygen	3700	73.56
Oldest Earth rocks	3800	75.54
Origin of the Earth	4600	91.45

Table 3a. Student Work Sheet/Biologic Events

Event	Millions of Years Ago	# of Mtrs from Closest Goal
Anatomically Modern Humans (*Homo sapiens*)	0.05	
Early *Homo sapiens* develop	0.3	
Development of *Homo erectus*	1.2	
Australopithecines and *Homo habilis* develop	3.2	
Development of Early Primates	35	
Extinction of the dinosaurs – "Great Extinction"	65	
Flowering Plants develop	140	
Dinosaurs are abundant	175	
First birds	180	
First mammals	220	
First dinosaurs	235	
Rapid expansion of living things – "Permian Explosion"	250	
First reptiles	300	
Development of the self–contained egg	340	
Tree appear	350	
First amphibians	360	
Insect–like creatures appear	400	
Earliest fishes	500	
Early shelled organisms	570	
Marine invertebrates abundant	600	
Multicelled plants and animals	700	
Advanced single cells	1000	
Development of eukaryotic cells	1400	
Early algae (blue–green) Gunflint formation	2200	
First life (single celled prokaryotes)	3500	

Table 3b. Student Work Sheet/Biologic Events — Answer Key

Event	Millions of Years Ago	# of Mtrs from Closest Goal
Anatomically Modern Humans (*Homo sapiens*)	0.05	0.001
Early *Homo sapiens* develop	0.3	0.006
Development of *Homo erectus*	1.2	0.01
Australopithecines and *Homo habilis* develop	3.2	0.06
Development of Early Primates	35	1.29
Extinction of the dinosaurs – "Great Extinction"	65	1.29
Flowering Plants develop	140	2.78
Dinosaurs are abundant	175	3.48
First birds	180	3.58
First mammals	220	4.37
First dinosaurs	235	4.67
Rapid expansion of living things – "Permian Explosion"	250	4.97
First reptiles	300	5.96
Development of the self–contained egg	340	6.76
Tree appear	350	6.96
First amphibians	360	7.16
Insect–like creatures appear	400	7.95
Earliest fishes	500	9.94
Early shelled organisms	570	11.34
Marine invertebrates abundant	600	11.93
Multicelled plants and animals	700	13.92
Advanced single cells	1000	19.88
Development of eukaryotic cells	1400	27.83
Early algae (blue–green) Gunflint formation	2200	43.74
First life (single celled prokaryotes)	3500	69.58

ILLUSTRATING THE PRINCIPLE OF THE FILTER BRIDGE

Based on an original activity by
Robert E. Collins and Richard W. Olsen

An explanation of the effect of competition and barriers on geographic distribution is provided. Also discussed are examples of barriers to dispersal and filter bridges that permit selected species to migrate through.

Two sterile petri dishes are placed side by side and joined by a bridge connecting both. One dish contains agar medium with sugar, while the other has a nonsugar medium. A mold, *Aspergillus niger,* and a bacterium, *Bacillus cereus,* are inoculated onto the sugar–agar medium. In time, the mold will crowd out the growth of the bacterium because it is more highly adapted for life in that environment. The bacteria will move through the filter bridge onto the nonsugar medium because it can use resources that the mold cannot.

Evolutionary Principles Illustrated

- Geographic distribution
- Filter bridges
- Competitive exclusion

Introduction

Throughout time, species dispersal has been affected by barriers and what have been called "filter bridges." These barriers and filters may be either mechanical or ecological. Mechanical barriers preventing dispersal of species are salt water, fresh water, deserts, jungles and mountain ranges. Ecological barriers could be lack of food, too many predators, or too much interspecies competition.

The geographical or ecological entity that represents a barrier for one species may be at the same time a filter bridge for another species. Filter bridges, in essence, are passageways that allow only some species to cross. For example, a desert is a mechanical barrier to a frog, a filter bridge to a camel, and perhaps no barrier at all to a bird.

Months or years are usually required to collect adequate field data to visualize patterns of species dispersal. The activity presented here can be prepared in a few hours and provides a living model using a bacterium and a fungus to illustrate the effects of competition and barriers on geographical distribution.

Intended Audience

- General biology
- Advanced biology

Materials (for each student group)

- 8 aluminum or enameled pans (34 cm x 25 cm x 5 cm)
- 8 Pyrex® glass plates (cut to cover pans)
- 4 Erlenmeyer flasks (500 ml)
- 18 sterile, disposable petri dishes (15 mm x 100 mm)
- stainless steel spatula
- Bunsen burner

- masking tape (2.5-cm wide)
- inoculating loop
- agar (8 g)
- peptone (2.5 g)
- maltose (20 g)
- beef extract (15 g)

Materials (to be shared by all class members)

- *Aspergillus niger* culture
- *Bacillus cereus* culture
- autoclave

Procedure

Two media are required. Prepare Medium A by combining 19 g agar, 5 g peptone, 20 g maltose, and enough water to make 500 ml. Heat to boiling to dissolve ingredients completely. Transfer equal volumes to two (500 ml) Erlenmeyer flasks and autoclave for 15 minutes at 15 lbs. to sterilize.

Prepare Medium B in the same manner by combining 8 g agar, 2.5 g peptone, 15 g beef extract, and enough water to make 500 ml. After autoclaving both media, cool slightly. Prepare petri plates by filling bottom halves to near capacity using sterile technique. Prepare a minimum of eight plates of Medium A and 12 plates of Medium B. Store agar plates in the refrigerator until ready for inoculation.

Prepare aluminum pans and glass covers for use by autoclaving for 15 minutes. Allow autoclave to cool before opening. As the pans and glass covers are removed, seal a glass cover to the top of each pan with masking tape to help maintain sterile conditions. Store pans at room temperature until ready for inoculation.

Sterilize the spatula in a Bunsen burner flame. Use sterile technique to transfer solidified agar from petri plates to a sterile pan and position as shown in Figure 1. With a sterile spatula, cut a 2x4-cm bridge of Medium B and position it

tightly between the agar plates as shown in the figure. Two bridges can be taken from one plate.

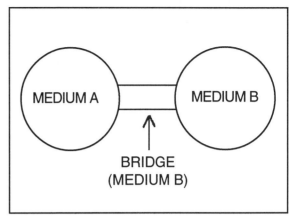

Fig. 1. Agar plates are positioned in an aluminum pan to construct the model.

Inoculate Medium A in the pan with *Aspergillus* agar approximately 2.5 cm below the upper border (12 o'clock, when plate is viewed as a clock face) and with *Bacillus cereus* approximately 2.5 cm above the lower border (6 o'clock) of Medium A. Seal the glass plate to the pan and incubate in the dark at 25° C.

Discussion

Within two days after inoculation, both the mold and bacteria should show a distinct growth on Medium A. After four days, the mold begins to crowd out the bacteria. By the 10th day, the bacteria have crossed the bridge and are establishing themselves on Medium B. After 14 days, only mold is observable on Medium A and only bacteria on Medium B.

This growth progression is best observed on a daily basis; however, daily observations are not always possible. To circumvent this or to accommodate student observations over a five day per week schedule, compare and inoculate eight pans on a staggered schedule and display three pans each day to demonstrate key stages in the growth progression. It is possible to get the stages needed for a five-day display schedule by inoculating one pan 18, 14, 12, 6, 5, 4, 3, and 2 days prior to the last day of the intended display period. It is useful to label the glass pan covers and display mounted pictures to clarify the principle (Figure 2 below).

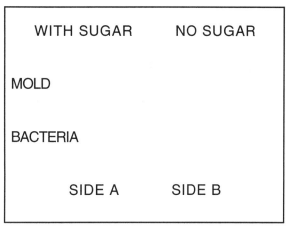

Fig. 2. The glass pan covers are labeled to clarify the model.

The model may represent two continents and a connecting bridge, such as North and South America with Central America between. The mold has competitive advantage on "Continent A," which is a sugar (maltose) medium; its long hyphae crowd out the bacteria and deprive them of needed nutrients. The bacteria, however, are able to survive because they escape across the protein (beef extract) medium bridge and establish themselves on "Continent B." Sugar is essential for mold's growth but is not essential to the growth of the bacteria. Thus, the bridge of protein medium is a barrier to the dispersal of the mold but is a filter bridge to the survival and dispersal of the bacteria.

With this model, students should have less difficulty understanding the idea that a species that is the most competitive in one environment may be the least competitive in another. In addition, a species that is the least competitive in one environment may be the most competitive in another; that is, the species most fit for survival (or dispersal) is always evaluated with reference to a particular environment.

Reference

This activity is based on an exercise by R.E. Collins and R.W. Olsen (1974). Filter bridges illustrate an evolutionary principle. *The American Biology Teacher, 36*(8), 474-475, 511, and is modified and reprinted with permission of the publisher.

IV. VARIATION WITHIN THE SPECIES

This chapter features several activities designed to help students understand that variation within the species is the raw material of evolution by natural selection. In spite of the fact that neither Darwin nor Wallace could explain the source of the variation, both recognized that something in the physical, behavioral or biochemical design of some members of a population permitted success; while others failed to survive.

Darwin viewed the variation within the population to be so important that he included an extensive evaluation of the diversity found in pigeons in his book, *The Origin of the Species*. Darwin reasoned that, if pigeon breeders could produce countless varieties intentionally in just a few years by selecting desirable traits from the normal variation within the pigeon population, nature could be capable of almost anything over millennia.

Activities here target the basic concept that, although there is normal variation within natural populations, sometimes one must look for it. The first exercise takes place at the zoo, where students examine groups of animals for subtle variations while they hypothesize about why some groups of traits are common in animals living in the same environment.

A second activity provides instructions for students to examine variation at a higher level by studying different, closely-related species of fruit flies. The unique aspect of this procedure is that students are able to gauge the amount of variation after speciation has already occurred.

The final activity in this section provides several examples of variation to investigate in the human species, for example, as humans are not permitted or denied reproduction based solely on their sets of individual traits.

ILLUSTRATING VARIATION AND ADAPTATION AT THE ZOO

Based on an original activity by
William F. McComas

During a visit to the zoo, students investigate variation within a species and have an opportunity to visualize the link between physical characteristics possessed by organisms and the environment they inhabit.

Evolutionary Principles Illustrated

- Variation within a species
- Adaptations
- Convergent evolution
- Predicting future evolutionary trends

Introduction

The accompanying work sheets each target a major principle pertaining to evolution by natural selection. The work sheets have not been designed to discuss fully or explain evolutionary theory but to foster classroom discussion by providing exciting illustrations of the principles involved. The individual sections may be modified for specific student groups or may be used independently of each other.

One major consideration relative to the potential success of this activity is the size and diversity of the collection at the zoo itself. It is very important that the zoo have a large variety of animals, preferably from different geographic areas, with at least three representatives of each animal to be investigated. This may become easier with the new emphasis in zoos toward having more examples of each species but fewer total species.

For the first activity, it would be possible to use farm animals or the resources of a large local pet store or animal shelter. This laboratory may also be performed with plants found at an arboretum or large nursery to illustrate that evolution is a unifying concept in all of biology, not just in the animal kingdom.

Students should be reminded that this activity primarily illustrates aspects of the Darwinian view of evolution by natural selection but that competing or modifying views — such as punctuated equilibrium — do exist. Despite the fact that science is still trying to establish the best explanation for the process of change through time, no one seriously questions the notion that change has indeed occurred.

Intended Audience

- Appropriate for all students

Materials (for each student group)

- photocopy of each work sheet (see pp. 48-50)

Procedure

Part I: Variation Within a Species

This first section is designed to illustrate the principle of variation within a species. Students are given the opportunity to study a particular type of organism and establish that even organisms of the same species will show some variation within that species. Generally, physical characteristics are considered by students to be the principal factors that enable some individuals to survive, but there are also differences in physiology, behavior and reproductive success. Although these factors may not be as apparent to the casual observer, they play just as big a role in the process.

Each student should choose a different animal species on exhibit at the zoo. These animals are studied as a group; therefore, several representatives of the animal should be available. A quick look at the individual animals may reveal nothing special, but a more detailed examination of many individuals will quickly reveal subtle differences among them.

Attention should be paid in this section to sexual differences within a species. The students should be aware that, in those species where such differences are common, they should examine a variety of animals of the same species and of the same sex. Sexual differences can, of course, be an interesting discussion topic. Additional differences that might confuse students include those such as size and other age–related factors.

Part II: Adaptation to the Environment

Next, the students will examine in detail the environment in which the animal lives naturally. The student will try to see what characteristics seem to allow the animal to fit well into that setting. This exercise is not meant to imply that there is a single perfect form for a species in any situation. The form that finally arises as being best suited is defined in terms of the other types that are in competition within the population.

Part III: Convergent Evolution

In this section, students are asked to find another animal that shares a similar environment with the animal selected in the section addressing variation within a species. In this case, the environment can be a large physical environment (i.e., desert, forest, grassland, etc.) or a smaller scale nutritional environment (i.e., seed eaters versus fruit eaters in the same general area).

This section is designed to lead to an illustration of convergent and divergent evolution. In its most general form, convergent evolution is a trend that produces similarities between unrelated forms because they share a common environment. Whales and fish both have fin-like structures permitting movement within their watery environment, but this alone is not a suggestion that the two animals are closely related.

Pressure from the environment has dictated that, in order to be a contender within a given environment, animals living there must share many common tools. Organisms coexisting in a given environment will often have a number of characteristics in common as a result of the processes of change working on both populations at the same time. This concept of shared traits due to living in a similar environment is called *analogy*.

Some students may be lucky enough to have found two organisms that live in similar environments in different parts of the world. An example using animals that may be familiar to students would be the gerbil and the kangaroo rat. Both of these rodents live in the same type of desert environment, but the gerbil is native to the dry, sandy areas of Africa and Asia, while the kangaroo rat is found only in the deserts of North America. These animals resemble each other in a number of ways and have almost identical food requirements. If they lived in the same area, they would likely be in direct competition for resources.

It would be an interesting project to concentrate on the differences between the two rodents and see how each is fine tuned for its specific environment. This would take some research, but from this information it might be possible to predict which animal would survive if both lived together in either the southwest desert of North America or in Asia.

In another example, the student might choose the deer and the kangaroo. Both have very similar nutritional requirements and, upon close inspection, are seen to have almost identical construction of their skulls and teeth. The head of a kangaroo and the head of a deer are very much the same, not because of close relationship but as a result of convergent evolution. Deer, of course, are placental mammals and carry their young inside their bodies until birth, while the marsupial kangaroos raise their young primarily in an external pouch. There are many other examples of placental mammals and their marsupial counterparts that students may discover.

Examples such as these show why an introduced species may effectively outcompete the native occupant of an environment. The transplanted or introduced organism may find itself well suited for the new environment but lacking the population controls provided by its own natural enemies.

Part IV: Predicting Future Evolutionary Trends

Finally, the students are asked to imagine what changes evolution might cause in a group of animals if their present environment slowly changed. This part of the exercise represents a simplistic view of a complex process but is useful in encouraging students to apply what they have learned. Furthermore, it may be

possible to discern more about the students' understanding of evolution from this creative exercise than any number of typical objective test questions.

Students should keep in mind that, even if a small group of individuals in a population possesses an advantageous characteristic, this characteristic may not be easily passed on to the next generation. This concept of heritability is poorly understood but is central to any discussion of future evolutionary ends.

With these cautions in mind, we could suppose that a desert environment becomes a woodland over a long period of time. Students might logically predict that desert toads which already have a slight green coloration might be "selected" by the environment since they would be more effectively hidden in the green of the forest. At the same time, tan-colored animals might not be able to hide as effectively and would be removed by predators.

Teachers may find it useful to consult the book *After Man* by Dixon (1981). This fascinating book makes predictions of what future creatures may look like based upon the present evolutionary trends coupled with projections of the pattern of continental drift.

Reference

This activity is based on an earlier exercise by W.F. McComas (1988). Variation, adaptation and evolution at the zoo. *The American Biology Teacher, 50*(6), 379–383, and is modified and reprinted with permission of the publisher.

Evidences of Variation, Adaptation and Evolution at the Zoo
Student Work Sheet

Part I — Variation Within a Species

1. Choose an animal in the zoo that is represented by at least three (3) different individual specimens. WRITE the common and scientific names of the animals on the lines below:

Animal A

_____   _____
common name   scientific name

2. What is the specific location of the animal in the zoo? _____

3. Examine your animals in detail and LIST as many individual differences as possible for the species in question. Example: Hair (long, short, or medium); Light brown vs. Dark brown fur color, etc.

   <u>Characteristic</u>   <u>Variation seen within the species</u>
a.
b.
c.
d.
e.
f.

4. DISCUSS the role of variation within a species in the process of evolution by natural selection.

Part II — Adaptation to the Environment

In this section, you are to examine the environment in which the animal you have chosen naturally lives. Try to see what general characteristics make the animal fit well into that setting and suggest what other characteristics, if present, would make the animal less well adapted to that particular environment.

5. WRITE a short paragraph which discusses the environment in which the type of animal you have chosen lives. Be very specific. Note — you may need to do some additional research here!

6. SELECT and LIST those characteristics which you believe will help the animal fits into its environment. Example: Long fur to help the animal stay warm in cold temperatures, etc.

   <u>Characteristic</u>   <u>Why does the characteristic help the animal fit into its environment?</u>
a.
b.
c.
d.
e.
f.

7. LIST a few general characteristics that would make the animal poorly suited to its normal environment.

       <u>Characteristic</u>   <u>Why would this characteristic be harmful?</u>

a.

b.

c.

8. In the space below, DISCUSS the advantages and disadvantages of having particular characteristics in relation to the process of evolution by natural selection.

Part IIIA — Convergent Evolution Due to a Shared Environment

In this section, you are to find another animal that lives in the same type of environment as Animal A. Note — The two animals chosen may both live in the desert, but they do not necessarily have to live in the same desert.

9. CHOOSE a new animal which lives in the SAME type of environment as "Animal A." WRITE the common and scientific names of the animal on the lines below:

Animal B

_____   _____

common name   scientific name

10. What is the specific location of the animal in the zoo? _____

11. EXAMINE "Animal B" and LIST the characteristics that it has in common with the organism you chose at the beginning (Animal A).

Both animals have:

a.   d.

b.   e.

c.   f.

12. DEFINE and DISCUSS the term **Convergent Evolution**.

13. Why do you think two animals that share a common environment have so many characteristics in common? How could this similarity have occurred?

14. What do you think might happen if the two animals you have identified lived in not only the same type of environment but also in the same area?

Part IIIB — Divergent Evolution Due to Geographic Separation

15. To further illustrate the idea of *convergent evolution,* IDENTIFY yet another animal that shares a high percentage of the same characteristics with "Animal A." WRITE the common and scientific names of this new animal on the lines on the next page:

Animal C

_____   _____
common name   scientific name

16. What is the specific location of the animal in the zoo?_____

17. EXAMINE this animal and LIST the characteristics that it shares with the organism you chose at the beginning.

a.   d.
b.   e.
c.   f.

18. STATE and DISCUSS the relationships, if any, between the two organisms that you have identified. Relationships include predator-prey, competitors, helpers (mutualism/symbiosis), amensalism (no relationship), etc.

19. DEFINE and DISCUSS the term **Divergent Evolution**.

Part IV — Future Evolution

Finally, go back to the animal you chose first (Animal A) and try to imagine what evolution would do to this animal if its environment slowly changed toward one quite different from that seen at present. For example, you might examine an animal in a desert environment and predict what would happen to it evolutionarily if the environment slowly became more like a woodland.

20. RESTATE the common name of "Animal A" and the type of environment in which it now lives.

21. Suppose that for some reason the animal's normal environment slowly changes, CHOOSE a new environment into which the old one will change. DESCRIBE this new environment. Specifically, what will be different about it?

22. Choose eight (8) characteristics seen in the animal at present and show how those characteristics will have to change (if they must) as the environment changes in order for the species to survive. Example, if a woodland slowly becomes a desert, the green coloration of a species toad might shift to brown so that it could hide more effectively.

Present Characteristic	Future Characteristic	Reason
a.		
b.		
c.		
d.		
e.		
f.		
g.		
h.		

23. Do you think that the animal in question will be able to live in the new environment proposed for it? To help you answer the question, think about the normal variation within the species. Do any of the individuals that you have observed have any of the characteristics that would enable it to survive and reproduce as the environment changes?

A SPECIES APPROACH TO EVOLUTION EDUCATION

Based on an original activity by
Dorothy B. Rosenthal

One approach to studying evolution is the investigation of species within a single genus. Through this approach, students can observe and appreciate the tiny but real differences that are the raw material of evolution by natural selection. This strategy focuses not only on the final results of evolution but also has the advantage of showing students the small, intermediate steps that must occur in the early stages of evolution by natural selection.

Evolutionary Principles Illustrated

- Diversity and variation
- Reproductive isolation

Introduction

Members of the genus *Drosophila* may be used to demonstrate a number of aspects of evolution theory. *Drosophila* are ideal for this purpose because many species are available and are easily reared in the laboratory. Through the species approach to evolution education students can learn that:

- Diversity is found at the genus level as well as at higher, more obvious taxons.

- Evolution occurs in small steps.

- Reproductive isolation is a significant factor in speciation.

- Species are usually, but not always, morphologically distinct.

- Organisms that do not interbreed in nature to produce fertile offspring are in different species, no matter how similar they may appear.

- Different species are adapted to different ecological niches.

This laboratory exercise enables students to learn worthwhile techniques, such as preparing insects for a collection, making permanent whole-mounts of insects, writing a species description, constructing a dichotomous key, and maintaining fruit flies under laboratory conditions. In addition, students may investigate the life cycle of a holometabolous insect, the morphology of insects in general, the importance of objective and quantitative descriptions of observations, and the significance of details related to those observations.

Intended Audience

- General biology
- Advanced biology

Materials (for each student group)

- cultures of *Drosophila virilis, D. melanogaster, D. mojavensis, D. pseudoobscura* and *D. persimilis*
- ethyl alcohol (10 ml)
- mounting pins (40)
- small fly–rearing vials (10)
- foam plugs

- *Drosophila* media, food and antimite paper
- xylene
- piccolyte mounting fluid

Procedure

The laboratory project outlined here is designed for a month of laboratory work by an advanced placement biology class (two 90–minute periods per week) or, with some modification, by students in a general biology course. Students are divided into teams of four, and each team is given cultures of one of the following species of *Drosophila*: *D. virilis*, *D. melanogaster*, *D. mojavensis* and *D. pseudoobscura* (and its sibling species, *D. persimilis*). Cultures may be obtained from scientific supply companies.

The stock cultures should be maintained by the instructor, but each group should be responsible for maintaining cultures of their own species and observing stages in the life cycle. Cultures are kept at room temperature in ambient light in plastic vials with foam plugs and antimite paper using standard techniques.

Initially, all five species are just "fruit flies" to the students, but as the flies are examined more closely, students will find that all, except *D. pseudoobscura* and *D. persimilis*, are quite distinctive. *Drosophila pseudoobscura* and *D. persimilis* are sibling species, indistinguishable on the basic of gross morphology alone. The differences to be noted in these sibling species include indistinct aspects of behavior, chromosome arrangement and habitat (Dobzhansky & Epling 1944; Prakash 1977). The process of learning to recognize the four morphological groups of flies is a valuable experience in observation.

After the students learn to recognize the flies easily, they should be given one or more of the following assignments:

- Preservation of specimens for later study.

- Description of their own species, using accepted scientific terminology and quantitative characters whenever possible.

- Development of a key for the species under study.

- Experimentation with breeding and competition.

A) Preservation of Specimens

Students are asked to preserve their specimens using an appropriate method, such as on dry insect mounts, storage in 75% alcohol, or in permanent whole mounts on microscope slides.

- For the dry mounts, flies dispatched in a killing jar are mounted on entomological "points," labeled, and pinned to the bottom of the box.

- Specimens are easily preserved in 75% alcohol, placed in tightly closed vials, and saved for future observation.

- Flies are prepared for permanent mounts by first dehydrating them in 95% and 5% alcohol and then clearing in xylene. The flies are then mounted in piccolyte, using regular or depression–type slides.

B) Species Descriptions

Each student is provided drawings and anatomical information on insects in general through sources such as Borror and White (1970) and for *Drosophila melanogaster* in particular (Sturtevant 1921). Students are also provided with an outline of the significant anatomical features of *D. melanogaster*. With this material and their preserved specimens, the students are then asked to write a description of their own species using scientific terminology, measurements and illustrations.

Although the work is somewhat painstaking,

students gain insight into both the degree of attention to detail and the rigor that is the basis for much scientific research. Because of the amount of work involved, team members found it useful to specialize in the different regions of the fly's body. Each team produced a paragraph or two describing its own species, along with a number of drawings. These descriptions could then be compared with standard descriptions of each species such as may be found in Sturtevant (1921), Patterson and Wheeler (1942), or Dobzhansky and Epling (1944).

C) Constructing a Key

In the process of describing its own species, each team will find it necessary to borrow specimens from the other groups for comparison. When all of the species descriptions are complete, it will be possible for the class as a whole to construct a dichotomous key to the species. Models of taxonomic keys are available in a number of sources.

D) Breeding and Competition Experiments

The experience the students gained in maintaining stock cultures is sufficient to permit them to carry out breeding and competition experiments. The breeding experiments consist of placing males of one species and females of a second species in a single culture vial (with the reciprocal cross in another vial) and observing the behavior and reproduction (or lack of parents from the sibling species, not their offspring [Dobzhansky & Epling 1944]). By contrast, mutant forms of one species *(D. melanogaster)* can be used to demonstrate that some obvious

differences, such as eye color, are not isolating mechanisms.

For the competition experiments, known numbers of two or more species are placed in a suitable *Drosophila* habitat and allowed to reproduce. At the end of the experiment, reproductive success can be measured by counting the number of each species.

A similar study of evolution within a genus has been described for "wind–bearing" desert forms of the genus *Haworthia*. Examples of these plants can be arranged to show a series of progressive modifications adapted to areas of drifting sands. Similar adaptations (convergence) are found in other genera of lilies, such as the *Aloe* and *Gasteria* (Newcomb, Gerloff & Whittingham 1964).

Author Acknowledgment

The author acknowledges the cooperation and enthusiasm of her AP Biology students at Sperry High School (Henrietta, NY). She is also indebted to Dr. David Wilcox, for discussions that originally led to the idea for this project, and to Dr. Ernst Caspari, for the numerous stimulating discussions on the subject of evolution.

Reference

These activities are modified from an original exercise by D.B. Rosenthal (1979). Using species of *Drosophila* to teach evolution. *The American Biology Teacher, 41*(9), 552–55, and are reprinted with permission of the publisher.

DEMONSTRATING VARIATION WITHIN THE SPECIES

Based on an original activity by
D.H. Keown

Through this exercise, biology students can realize the variability that exists in a population by carrying out several simple, fun classroom activities.

Evolutionary Principles Illustrated

• Genetic variation

Introduction

The definition of evolution is sometimes given as "change in the genetic makeup of a population over a period of time." This definition is not without criticism, for it is the total organism, with all of its systems and behaviors, that passes on the genetic code. Nonetheless, the definition is applicable to most organisms. It is the genetic blueprint that expresses the anatomy, physiology and behavior of the organism. Knowledge of the genetic material is important to an understanding of evolution.

Genetics instruction always precedes the explanation of evolutionary processes, but the ties that bind the mechanics of genetics to the process of evolution are often presented without proper emphasis. For instance, it is at the time of synapsis that the variability made possible by sexual reproduction is accomplished.

Synapsis is the time of mixing of the genes with the population's gene pool. Without this mixing, the variability of offspring is limited to mutations and chromosomal aberrations, the only

factors creating variability in asexual reproduction.

The argument can be made that Darwin discovered the mechanism of evolution without a knowledge of genetics, even before Mendel's laws were known. Though he was not familiar with genes, mutations or DNA, he was well aware of the variability of offspring in sexual reproduction. The activities presented here will help to exemplify the variation that is the material of evolution.

Intended Audience

• Life science
• General biology

Materials (for each student group)

• seeds for dihybrid crosses
• 5x8-inch index cards
• cloth tape measure (any length)

Procedure

The following ideas are provided as suggestions to help students visualize variation within the species. Teachers may wish to use one or more of these activities.

A) Students can propagate flowers, such as geraniums, vegetatively to see the lack of diversity in the offspring. They may also contrast this sameness with plants produced from crossing hybrids. Seeds from hybrid crosses are available from biological supply houses.

B) Analysis of a litter of kittens or puppies illustrates the variability in mammals quite well. Students may analyze "personality" characteristics — height, weight, color, tail length and behavior, for example.

C) Humans may also be used to illustrate variation within the species, although it is important to point out that survival of individual humans is determined much less by these physical characteristics than for organisms exposed to pressures in the natural world.

- The variability of features expressed by a population of humans in a biology class is interesting and may show the nonuniformity that natural selection might work upon if we were a natural existing species. Hand shapes can be analyzed by having the students trace the outlines of their hands on paper. Post the papers on the wall and compare them.

- Another activity shows reaction time among students. From a uniform height, drop a 5x8 file card between the outspread thumb and index finger of each student while he/she tries to grab the card. Make a mark on the card at the point where each student catches the card.

- Illustrate the varying ability to memorize by giving each student a verse of an obscure poem that none of the students has seen. Place it facedown on their desks and have them all turn it over at the same time and begin to memorize the verse. Tell them to raise their hands as soon as they have the verse memorized and record the times.

Care has to be exercised in these activities to see that students with slow reaction time or poor memorization skills are not embarrassed. One way to do this is to make the process a team event so that groups of two students work together to memorize the passages.

- You may also try using such features as head circumference, foot length, and other morphological features that are mainly genetically controlled to show the variation in the classroom population. With a large enough sample, the classic bell-shaped curve will result if the data for one of these characteristics are graphed.

Reference

These activities are modified from an article by D. Keown (1988). Teaching evolution: Improved approaches for unprepared students. *The American Biology Teacher, 50*(7), 407–410, and are modified and reprinted with permission of the publisher.

V. BIOTIC POTENTIAL AND SURVIVAL

D arwin and Wallace both read Thomas Malthus' book, *An Essay on the Principle of Population,* and referred to its central idea in their similar theories of evolution by natural selection. Malthus stated that "populations increase geometrically, but food supplies increase only arithmetically."

Although he was speaking about humans, Malthus made a valid point about all populations; they grow very quickly and soon outpace available food sources. From this conclusion, Wallace and Darwin inferred that natural populations would always produce offspring in excess of what the environment could support. As a consequence, there would be a struggle for survival that would be won only by the best suited individuals. Those individuals surviving would, of course, be the ones permitted by nature to reproduce, moving the traits that made them successful into the next generation.

Without overproduction, there would be no such competition and resulting natural selection. The activities presented in this section all demonstrate, either with simulation or by examination of real populations, the concepts of overproduction and competition.

THE ARITHMETIC OF EVOLUTION

Based on an original activity by
Roxie Esterle

These activities serve to acquaint students with the notion that organisms have the ability to reproduce vast numbers of offspring. This overproduction helps insure the survival of the species but also sets in motion a competition for survival between the offspring.

Evolutionary Principles Illustrated

• Overproduction
• Geometric growth of species
• Biotic potential

Introduction

A major influence on both Darwin and Wallace was the work of Thomas Malthus, an English clergyman and economist. Malthus was able to show that the population of English colonies in America had doubled every 25 years from 1643 to 1760. Malthus concluded that there would be a "struggle for survival" between those who had sufficient resources and those who did not. In an amazing coincidence, Darwin and Wallace both applied this idea of a "struggle for survival" to their theories of evolution by natural selection.

Intended Audience

• Appropriate for all students

Materials (for each student group)

• world population data table
• "Arithmetic of Evolution" data table (p. 62)
• graph paper (semi-logarithmic optional)
• various types of plants and fruits for dissection
• natural history reference books

Procedure

The examples provided below are all good ways to illustrate the concept of overproduction. Teachers should use the most appropriate examples with their students.

Part I: Species Growth Potential in Plants

Have students dissect an assortment of plants and count the number of seeds, on average, in each. Examples might include seeds from apples, tomatoes, pears or pine cones. Using an appropriate reference book, students can determine the average number of fruits per adult plant. If one assumes that all of the seeds from a given tree germinate, it is a simple task to calculate how many offspring will be produced by a single plant in one growing season. Of course, one can then calculate how many offspring a given plant will produce in a lifetime.

Botanical Examples for Discussion (Otto & Towle 1973, p. 207):

• A single fern plant produces 50,000,000 spores per year.

• A mustard plant produces 730,000 seeds. If they all matured, the adult plants would cover an area 2000 times the land surface of the Earth in just two years.

Part II: Species Growth Potential in Animals

Using an appropriate reference book, students can compare the average litter size of various animals. As in the plant activity, if one assumes that all of the offspring survive and live indefinitely, the total number of offspring per adult can be determined. For example, students can calculate mouse population growth based on four offspring per litter, 21 days of gestation, 21 days to sexual maturity, and a lifespan of about one year.

Zoological Examples for Discussion (adapted from Otto & Towle 1973, p. 207):

• An oyster produces 114,000,000 eggs at a single spawning. In five generations, there would be more oysters than the estimated number of elections in the visible universe.

• Although an elephant produces only six young per lifetime, if all of these offspring lived, in 750 years 19,000,000 elephants would be produced from the first mated pair.

• A sea hare (a marine annelid worm) produces 14 billion eggs during its lifetime. If all hatch and mature, the Earth would be many feet deep in sea hares in a few generations. In actuality, only five offspring from each generation ever reach maturity.

Part III: Human Population Growth

On a piece of graph paper, label the X axis "Time" and the Y axis "Population" and graph the population data shown in Table 1 in the next column.

If students are asked to graph these data on standard graph paper, a "J" shaped curve will result. This type of growth pattern is typical in populations increasing at a logarithmic rate. It might be interesting to have students graph the same data on semi-logarithmic paper. In doing so, they will produce a straight line because one axis of the graph marks changes at a rate 10 times the others.

Table 1. Estimated World Population.

Year	Population
1750	760,000,000
1760	803,000,000
1770	848,000,000
1780	896,000,000
1790	947,000,000
1800	1,000,000,000
1810	1,039,000,000
1820	1,080,000,000
1830	1,122,000,000
1840	1,165,000,000
1850	1,211,000,000
1860	1,258,000,000
1870	1,363,000,000
1890	1,534,000,000
1900	1,628,000,000
1910	1,741,000,000
1920	1,861,000,000
1930	2,070,000,000
1940	2,296,000,000
1950	2,517,000,000
1960	3,019,000,000
1970	3,698,000,000
1980	4,448,000,000
1990	5,292,000,000
2000	6,261,000,000

Sample Discussion Questions

1. If the population of the United States continues to grow at the present rate, what will be the total in the year 2000?

2. What resources in the United States are likely to limit population growth?

3. What factors have influenced the population growth in the United States? List these factors and explain the influence of each factor.

4. Is "natural selection" in the Wallace/Darwin sense working in any human populations? Provide support for your answer.

5. Has our birth rate changed? If so, why?

Reference

This previously unpublished activity, titled *The Arithmetic of Evolution,* was contributed by Roxie Esterle, a science consultant specializing in evolution education.

"The Arithmetic of Evolution"
Data Sheet

Plants

1. Name of the flower or plant. ——————————————————

2. Locate the number of seeds and count the number per flower or plant. ——————

3. Count or estimate the number of flowers per plant. ——————————

4. How many times per year does the plant produce seed? ——————————

5. What is the total number of potential offspring per year? ——————————
(Space for calculations)

6. If all of the **offspring** were to survive and reproduce **one** time. approximately how many would there be? ——————————————————————————
(Space for calculations)

Animals

1. Name of animal. ——————————————————————

2. Number of offspring per litter. ——————————————————

3. Number of litters per lifetime. ——————————————————

4. Generation time (time from birth to reproductive maturity). ——————————

5. If all of the **offspring** were to survive and reproduce **one** time, approximately how many would there be? ——————————————————————
(Space for calculations)

DEMONSTRATING BIOTIC POTENTIAL

Based on an original activity by
D.H. Keown

Students use fishing worms (red wigglers) to study popoulation dynamics. Students can count and graph worm numbers on a daily basis to visualize population growth.

Evolutionary Principles Illustrated

- Competition
- Overproduction

Introduction

This activity will graphically show students the potential for species overpopulation. Because of this potential in natural populations, most of the offspring do not survive to reproductive age, and the "struggle for survival" that became the focal point of Darwin's and Wallace's discovery results.

Intended Audience

- Life science
- General biology

Materials (for each student group)

- 4 red wigglers (purchased from a bait dealer)
- quart container filled with peat moss and rich soil
- box of Total™ or Wheaties™ (to be used as worm food)

Procedure

The concepts of biotic potential are illustrated concretely by culturing some fast-reproducing organisms in a closed environment. Commercial fishing worms called "red wigglers," purchased from bait dealers, may be used for this purpose. Place four worms in quart-sized cottage cheese cartons filled almost to the top with a media of peat moss and rich soil. A finely ground break-fast food provides a good diet for the worms. The cereal is sprinkled on top of the media and the culture is kept moist and cool.

At two-week intervals, have the students count, record, and graph (time against population size) the number of worms in the cartons. Also, have the students measure the average size of 10 worms at each counting, since there is another serendipitous outcome for the students to see.

As the population peaks, the worms begin to diminish in size and number until there are none left. The students see a real example of the earthworm's biotic potential and a stark example of the result of overcrowding, resulting in depletion of resources and the accumulation of waste.

Reference

This activity is modified and reprinted from an activity by D.H. Keown (1988). Teaching evolution: Improved approaches for unprepared students. *The American Biology Teacher, 50*(7), 407–410, with permission of the publisher.

SIMULATING POPULATION DYNAMICS

An original activity by
Brian J. Alters

This activity will model such concepts as carrying capacity, exponential growth, distribution over time, zero population growth and possible extinction. Some effects of chance, immigration, emigration, competition, disease, pollution and seasonal changes are also included in this model.

Student groups run probability experiments in which inanimate populations grow and level off due to multiple factors; after which, students construct population curves that are different for almost every population (student group) due to chance. This is accomplished through the use of dice and plastic or paper chips (other items may be used instead of chips).

Evolutionary Principles Illustrated

- Struggle for survival
- Extinction
- The role of chance in evolution

Introduction

The first two activities in this chapter demonstrate to students that organisms have the ability to reproduce vast numbers of offspring when conditions are ideal — such as abundant food and living space, no organisms competing for those resources, and no predators or disease present. This biotic potential is rarely realized in nature.

Most organisms do not survive to reproduce fully or reproduce at all due to the "struggle for survival." This concept became the focal point of Darwin's and Wallace's discovery. The activity will symbolically and graphically allow students to experience the restrictive variables on overpopulation.

Intended Audience

- General biology
- Advanced biology

Materials (for each student group)

- 1 pair of dice
- 100 paper chips of any color (or virtually anything small with which to keep tallies)
- 200 paper chips of another color (or anything with which to keep tallies)
- 1 sheet of graph paper (optional)

Procedure

Each student group should have a pair of dice, 100 of one colored item and 200 of another colored item (for example, 100 red chips and 200 green chips). The dice introduce the element of chance. Each red chip represents an individual organism that potentially can reproduce.

In population activities, only the potentially reproducing organisms "count," so there is no need to mention the nonreproducing sex, if any. Each green chip represents the resources necessary (consumed) per individual per year.

Each individual has the capability of reproducing once a year. A year is defined as the duration to role the dice once for each individual already in the population for that year. For example, if the population has five individuals and the third roll of the dice reads "4," meaning "one immigration," (dice values are on the Student Niche handout) the population would still only have two more rolls for that year, even though another individual joined the population for that year (it will have a chance to roll the dice next year).

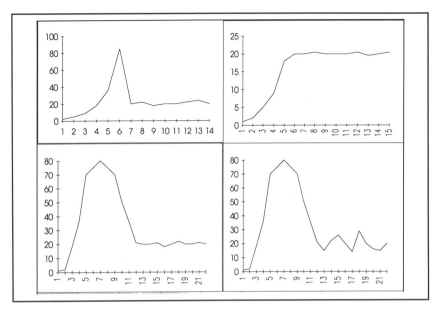

Fig. 1. Graph possibilities.

At the beginning of each year, the population receives 20 resource chips. During each year, each individual will use up one resource chip, to be given up at the end of the year (after all rolls have been made for the year). And, as above, if an individual newly joined the population that year, then there is no need to give up a resource chip (it will eat and drink next year).

If there are not enough resource chips for each individual (not counting the newly joined or born individuals), then those individuals without resource chips, leave the population (die and, to a much lesser extent, emigrate).

Most Important: If an individual joins the population within the year, it is not considered when rolling dice or exhausting resource chips for that year. However, it is counted as part of the total population count at the end of the year. *(Note: There is a stepwise summary of the above procedure on the "Student Niche" page.)*

Figure 1 represents some possible graphs for this exercise (years versus population). Notice that, in each case, a carrying capacity was established. Although each student group will have a different graph for its population, each probably will have established a carrying capacity (except for extinctions).

After completion of the activity, the teacher could lead a discussion with the entire class to share what other student group populations experienced and to discuss some topics such as biotic potential, extinctions, and how they relate to the survival of the fittest. The next page is to be photocopied and handed out to the student groups as their instructions for the exercise.

Reference

This previously unpublished activity was contributed by Brian J. Alters, a Ph.D. candidate in science education at the University of Southern California, Los Angeles, California.

Population Dynamics
Student Niche

First Year

1. Start with 5 individuals in the population and 20 resource chips.

2. Roll the dice once for each of the 5 individuals. Move individual and resource chips in and out of the population depending on what is rolled on the dice:

If the total on the dice is:

    2, then forfeit 5 resource chips (due to a nonfavorable seasonal change)
    3, then record one birth
    4, then record one immigration
    5, then record one birth
    6, then record one birth
    7, then record one birth
    8, then record one death (due to competition)
    9, then record one death (due to disease)
  10, then record one emigration
  11, then record one death (due to pollution)
  12, then receive 5 resource chips (favorable seasonal change)

3. After the five rolls, the year has ended. Take away one resource chip for each of the individuals that were in the population at the beginning of the year (newcomers don't eat or drink until next year).

4. Count the total number of individuals now in the population. Record the number and the year on paper. You are now ready to begin the following years.

Following Years

5. Give your population 20 resource chips.

6. Roll the dice once for each of the individuals that are in the population at the <u>beginning of this year</u> (newcomers [births and immigrations] will roll next year). Move individual and resource chips in and out of the population depending on what is rolled on the dice, as above.

7. After the dice have been rolled once for each individual that <u>started the year</u>, the year has ended. Take away one resource chip for each of the individuals that were in the population at the <u>beginning of the year</u> (newcomers don't eat or drink until next year). If there are more individuals that resource chips, those individuals die or, to a lesser extent, emigrate out of the population.

8. Count the total number of individuals now in the population. Record the number and the year on paper.

9. Repeat steps 5-9 until the population has struggled through 20 years.

VI. ADAPTATION

Adaptation is a biological term that is universally misunderstood by students. In common language, the term represents something over which an individual has control. Generally, people are said to be able to adapt to or change.

In the world of nonhuman living things, however, the term "adaptation" means a characteristic or set of characteristics *already possessed* by an individual, giving it an advantage over others in the struggle for survival. Organisms cannot choose to adapt; they must already possess the raw materials for such adaptation. Hence, the term "adaptation" used in biology is a noun, not a verb.

Students will be able to see a number of traits and judge the adaptive value of such traits in the activities provided in this chapter. Using a variety of instruments, such as drinking straws, tweezers and pliers, students are asked to pick up various types of seeds in a simulation of bird beaks. Obviously, some "beaks" will be better adapted than others for certain kinds of seeds.

In another experiment, real birds are employed to pick up food that has been dyed different colors. This is an authentic test of the birds' abilities to see various food sources against a colored background. Success in food gathering is good news for the bird — but bad news if the prey items are plant seeds or insect larvae.

MODELING THE PRINCIPLES OF ADAPTIVE RADIATION

Based on an original activity by
Lawrence Blackbeer, Arthur P. Loring, and Kia K. Wang

This model illustrates the operation of the principles of adaptive radiation by modeling the process, with holes drilled in egg cartons and unsorted sand to represent species. Grains of sand that fall through the holes in the egg carton compartments are said to lack the traits necessary to survive.

Evolutionary Principles Illustrated

* Geographic isolation
* Competition
* Natural selection
* Speciation
* Rapid environmental change

Introduction

The principles of adaptive radiation have operated throughout geologic time. The taxonomic categories that biologists apply to flora and fauna actually represent the links in a continuous chain formed as a result of these principles.

Intended Audience

* General biology
* Advanced biology

Materials (for each student group)

* egg carton
* drill with bits of various sizes (e.g., 0.75 mm, 1.25 mm, 1.75 mm, 2.5 mm)
* approximately 500 g of poorly sorted quartz sand
* 35-mesh sieve

Procedure

The basic apparatus consists of the six compartments in half an egg carton. Each compartment is numbered, consecutively, for reference purposes. Each compartment represents the geographic region in which a different species functions. The carton as a whole represents a much larger environment such as an ocean or a lake. The vertical and horizontal variability within each compartment reflects the variable environmental conditions within the species' adaptive zones.

Nine holes equally spaced in three columns are drilled in the bottom of each compartment. The holes are meant to illustrate the sievelike operation of natural selection. The size of the holes varies from compartment to compartment. This variability of the hole diameters is intended to reflect the variable environmental conditions within geographic niches one through six. The hole diameters for Compartments one through six are as follows:

1. 0.75 mm
2. 1.25 mm
3. 1.75 mm
4. 2.50 mm
5. 2.50 mm
6. 2.50 mm.

The partitions separating the six compartments represent boundaries imposed by environmental conditions; that is, the next compartment is beyond the realm of a species' adaptability at that moment in its evolution. However, the boundaries are not entirely unbreachable, and preadapted organisms of one niche can gradually adapt to the environmental conditions of an adjacent niche.

To operate the model, one student will need about 500 g of poorly sorted quartz sand. The sand is separated by means of 57-mesh to 35-mesh sieves, with the result that only grains of 0.5 mm, 2 mm, and 4 mm in diameter are used in this model. These grains, deposited in a compartment, represent the individuals of a species. The different grain sizes reflect the normal phenotypic variability observed within living populations.

Initial Assumptions

The total environment is initially devoid of animal life. (This is not usually the situation in nature, but the validity of this procedure will be explained later). However, we must assume the plants are present to supply the food and oxygen that the animal occupants will require.

We assume, further, that the carton represents a single general environment — say, a freshwater lake. The grains that are to be poured into Compartment 1 represent members of a herbivorous species, A.

It is necessary to assume that the members of Species A have already adapted to slightly different environmental conditions elsewhere; then they have been transported, either artificially or accidentally, into this habitat (Compartment 1). They are sufficiently preadapted to function under the environmental conditions encountered in this habitat.

Procedure

PHASE I

Slowly pour the mixture of grains (from 0.5 mm to 2 mm in diameter) into Compartment 1. The grains that fall into this compartment are individuals of Species A. At the same time, gently shake the entire carton in order to sieve those grains that are smaller than the holes (.75 mm) in Compartment 1. The shaking also imparts a vertical sorting within each compartment. When Compartment 1 is filled with sediment, the first phase of this model has been completed.

This phase of the model illustrates the operation of intraspecific competition, natural selection and specialization. The pouring of grains and their immediate competition for space within the compartment reflect the mechanism of intraspecific competition. However, just as certain grains are sieved out of this compartment, natural selection effectively weeds out those members of a population that cannot adequately compete with the more competent individuals of the same species.

Those grains that fall onto the tray below represent the portion of the population that is relatively unfit and is removed from the niche by natural selection. Those grains remaining in Compartment 1 are to be considered specialized as a result of natural selection.

The criteria for direct selection has been shown to be the individual's phenotype; the frequencies of a species' alleles should, therefore, be indirectly affected by this process of selection.

Although not demonstrated in this phase, those genetic mutations manifest advantageous phenotype traits that would be incorporated into the species' gene pool via the natural selection process. It is important to note that all members of Species A experience natural selection under the environmental conditions within this habitat;

geographic isolation as a stimulus for divergence is negligible during Phase I.

When Compartment 1 is filled — that is, when the niche has reached its carrying capacity — the grains are distributed so that the larger grains accumulate at the top because of the shaking action. This zoning of the remains within the compartment is to be interpreted as the evolution of variations within a species by natural selection under slightly variable environmental conditions.

PHASE II

As the pouring of the grains continues beyond the capacity of Compartment 1, grains tumble over the partitions into the adjacent compartments. (Tilt the carton slightly so that grains will fall into Compartment 2. Grains falling into Compartments 3 and 4 are to be disregarded.)

As soon as the first grains tumble into Compartment 2, transfer the pouring of grains into that compartment only. Continue to shake the carton gently, to aid sieving. When the carrying capacity of Compartment 2 is reached it will show a smaller range in grain size than that of Compartment 1. This is because of the larger holes in Compartment 2. Analysis of the grains remaining in Compartment 2 reveals a range in diameter between 1.25 and 2 mm.

This phase of the model illustrates the operation of intraspecific competition, natural selection, and specialization, as discussed in Phase I. In addition, it is assumed that certain genetic mutations are being incorporated into the gene pool of the population within this habitat (Compartment 2).

At the beginning of Phase II, the factors of adaptive radiation have extended the limits of the adaptive zones of Species A by the formation of a new species within this habitat (Compartment 2). The assumption is that Habitat 1 is essentially isolated from Habitat 2. Therefore, in time, genetic mutation, intraspecific competition, natural selection, specialization, and geographic isolation should result in speciation.

Divergence in the model is said to have occurred once the carrying capacity of this habitat (Compartment 2) has been reached (at the end of Phase II). At this time, the grains within Compartment 2, which formerly represented a subspecies of Species A, have come to constitute a Species B — another herbivorous, aquatic, invertebrate stock. Thus, theoretically Species A and B can no longer hybridize (produce fertile offspring). But these species are still closely related genetically and phenotypically; they would be classified as members of the same genus.

PHASE III

The model thus far has shown how the pressures of natural selection, initiated by intraspecific competition and resulting in specialization, are resolved (Phase I). In addition, the continued operation of the principle of adaptive radiation has resulted in speciation — the evolution of Species B (Phase II). However, no pressure has yet been exerted as a result of intraspecific competition.

Phase III of the model illustrates the effect of interspecific competition between Species A and B. Continue to pour grains into Compartment 2 until the sand grains come into contact with the grains in Compartment 1 at the partition. The grains in each compartment actually "compete" with one another for the available space at the partition, where the niches of Species A and B overlap.

Because A and B are herbivorous, interspecific competition may occur for living space for food. In the overlapping region, A and B divide food and living space according to their relative degrees of adaptive competency. In nature, interspecific competition would tend to intensify intraspecific competition, thereby raising standards of fitness required of these species by their respective environmental factors.

Intraspecific competition also influences the direction in which divergence will occur. To show this, an unoccupied environmental space is considered to be an environment of low pressure, and a saturated habitat is considered to be an environment of high pressure. The continued pouring of grains into the saturated habitats of Species A and B (Compartments 1 and 2) causes grains to fall into vacant Compartments 4 and 3, respectively.

Note that the grains do not flow from one high–pressure environment (Compartment 1) into another high-pressure environment (Compartment 2). Instead, the grains flow from the saturated, or high pressure, habitats into the unsaturated, or low pressure, regions of the model. We suggest that the members of a species will tend to migrate into an environment of low pressure containing the proper unfilled niche. Habitats that contain niches already filled by competent species discourage the settling of similarly adapted species.

The larger hole-diameter of Compartment 3 weeds out grains that were previously permissible in Compartment 2. This occupance mimics the environmental selection in nature of a species' phenotypic potential and the indirect loss of certain alleles from a species' gene pool. Assuming that the individuals in this habitat (Compartment 3) are isolated geographically from those in the adjacent habitat (Compartment 2), genetic mutation, intraspecific competition, natural selection, and specialization will result in speciation.

Species C is said to have evolved in this habitat (Compartment 3) once the carrying capacity of this environment has been reached (at the end of Phase III). Species C, composed only of the

largest grains, is a third herbivorous, aquatic invertebrate species.

The grains poured into Compartment 1 overflow into Compartment 4. However, the holes in Compartment 4 are so large that all grains that fall into this compartment are sieved into the tray below. This suggests that the environmental conditions in Compartment 4 are beyond the realm of adaptability of Species A. This species cannot make the transition successfully, due to the nature of the environmental conditions encountered in the habitat (Compartment 4); no letter designation is therefore necessary. Because the holes in Compartments 5 and 6 are the same size as those in 4, they too will remain vacant (devoid of grains).

PHASE IV

After Compartments 1, 2 and 3 are filled with grains, a small nail is used to enlarge all their compartmental holes to 3.0 mm in diameter. After this is done, continue sieving. This illustrates the effect of rapid environmental change on intraspecific competition, natural selection and specialization.

Environments are constantly changing. They are dynamic systems. The organisms in these changing habitats may either adapt to these environmental conditions, migrate or become extinct. Environmental change is usually a slow process, during which natural selection modifies the phenotypic and genotypic norm of a species. However, if environmental change is relatively rapid, the effects may be drastic for the organisms within the habitats experiencing modification.

Discussion

Standards of fitness, determined by chemical, physical and biological environmental factors, are upset during rapid environmental change. The faunal inhabitants may not be suitably adapted to function under rapidly altered environmental conditions.

A genetic mutation is an accident. If a genetic mutation proves to be phenotypically advantageous during rapid environmental change, it may ultimately become the genetic norm of its species. Mutant alleles, which are usually recessive, may therefore become common due to the selective advantages of their phenotypic expressions. Thus, mutations are the safety valves of a species during periods of rapid environmental change.

This may be demonstrated in the model by adding 4.0 mm grains in diameter to Compartment 1 before Phase IV has been completed. These are not sieved. These unusually large grains represent the mutations of Species A that have been deemed fit by the new environmental conditions in this habitat (Compartment 1).

However, rapid environmental change may serve as an instrument of extinction of a species if no beneficial mutation occurs. Once Phase IV has been completed, Compartments 2 and 3 are devoid of grains. It is suggested that their faunal inhabitants — Species B and C, respectively — did not adapt or produce advantageous mutations during rapid environmental change.

While it might be useful to model factors such as intraspecific competition, natural selection, specialization, and divergence, this model is incapable of illustrating these modifying factors because of their dynamic nature. However, the teacher may wish to mention them to his/her students.

Validity of the Initial Assumption

The initial assumption of the model was that no faunal occupants existed in this environment before the introduction of Species A. During Phase I, interspecific competition was not occurring. Although this situation would not be representative of evolution within many existing environments, Species A could also have been introduced as a relatively competent species that was competing in Habitat 1 with several incompetent species (not identified).

In that case, the more competent Species A would have efficiently eliminated the less competent species. If the model had been introduced in this manner, the validity of the model and the conclusions would not have been altered. (A summary of the model's conclusions is illustrated in the accompanying figure.)

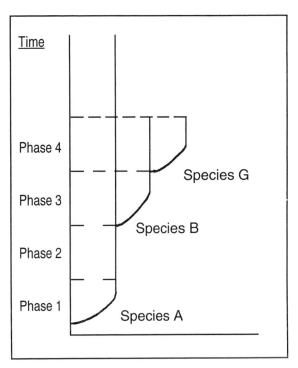

Fig. 1. Divergence of evolutionary events.

Reference

This activity is based on an original exercise by L. Blackbeer, A.P. Loring and K.W. Wang (1972). A teaching model of the principles of adaptive radiation. *The American Biology Teacher, 34*(8), 471–474 & 476, and is modified and reprinted with the permission of the publisher.

THE BIRDS AND THE BEAKS

Based on an original activity by
Roxie Esterle

This lab demonstrates the principles of natural selection by showing that different adaptations (usually physical structures) have value in a specific environment or for a specific purpose.

Evolutionary Principles Illustrated

- Adaptations
- Diversity
- Competition

Introduction

In this activity, students simulate the usefulness of various types of bird beaks by trying to pick up particular types of seeds with various "beak–like" tools. Such tools include pliers, knives, spoons, etc. It is possible to make this activity quite sophisticated by having students use many seeds and tools and then calculate the ratios of various seed types "captured" with particular tools. Conversely, this activity may also be used as a simple illustration of adaptive structures.

Intended Audience

- Life science
- General biology

Materials (for each student group)

- assorted tools of varying design, including pliers with different tip configurations
- two flat dishes (one will contain the mixed seed and one will contain the "eaten" seeds)
- four types of beans or seeds of varying sizes, such as sunflower seeds, kidney beans and flax seeds (A mixed bag of commercial bird seed will be useful.)

Procedure

1. Assemble an assortment of beans consisting of approximately one teaspoon of each type of seed mixed together for each student group.

2. Instruct students that they are to use their "tool" to pick up as many seeds as possible and put them into the now-empty flat dish.

3. Pliers will be used to represent the beaks of various birds, and the student using the pliers will represent the bird. Students should be instructed to use both hands to manipulate the pliers.

4. Each "bird" (student) is instructed to forage for seeds in the flat dish for one minute.

5. After one minute, all birds are asked to stop eating. Other members of the flock can sort and count beans and record their data in an appropriate chart.

6. Return all beans to the flat dish.

7. Repeat steps 4-6 as necessary until all the "birds" (students) in the laboratory group have had a chance to feed.

Name of Tool	Type of Seed Eaten				
Averages					

Table 1. The relationship between the number of seeds collected (eaten) by various tool types.

8. Depending upon the goals of the instructor or the nature of the learners, the data may be entered onto a table (see Table 1) or may simply be discussed qualitatively.

9. For those who would like to have students make graphs, the following relationships might prove illustrative:

• Seed type *versus* number of seeds eaten
• Tool type *versus* total number of seeds eaten
• Tool type *versus* weight of seeds eaten.

Suggested Discussion Questions

1. Which type of bird beak do you feel is the best adapted? Why?

2. Which type of seed do you feel is best adapted to avoid being eaten? Why?

3. Which bird beak (tool) functioned best as a "generalist"? This may be determined by looking at the data for your lab group. Did your bird beak catch an equal number of each type of seed or was it more successful with a specific type?

4. Which bird beak was most specialized?

5. What would happen to a bird in a natural situation if it was unable to secure an adequate number of seeds? What will happen to the bird that can catch the most seeds? (What will happen to the genes of each of these two birds?)

6. Are some of the "birds" in this activity more skilled than others in gathering seeds? Does this happen in nature?

7. Are all offspring from the same parents identical in their physical appearance? In their

ability level? Give some examples to support your answer.

8. Identify an example from real life experience where competition occurs among living things.

9. Do you think that all of the seeds are equal in nutritional value to the birds? Should this be a consideration in experiments such as this?

10. What factors influence exactly how much food a bird eats?

11. What are some of the sources of error in this activity?

12. How does this activity relate to the Darwin/Wallace explanation of how evolution occurs?

Reference

This previously unpublished activity, originally titled *All About the Birds and Beaks*, was contributed by Roxie Esterle, a science consultant specializing in evolution education.

DEMONSTRATING COLOR ADAPTATION IN FOOD SELECTION

Based on an original activity by Joseph Abruscato and Lois Kenney

In this activity, a bird is allowed to select and eat seeds that have been dyed either grey or brown and randomly scattered on a chessboard composed of gray and brown blocks. This exercise may be done as a demonstration or performed on successive days by student groups.

Evolutionary Principle Illustrated

- Adaptations (protective coloration)

Introduction

The way in which protective coloration ensures the survival of a species is a topic dealt with in most biology classes. Usually, examples of organisms displaying this evolutionary adaptation to the environment are presented to students through pictures or preserved specimens and samples from their environments. Using color adaptation, students discover this principle of survival.

Intended Audience

- General biology
- Advanced biology

Materials (for each student group)

- a tame bird of any species
- chessboard with brown and gray squares
- equal quantities of birdseed (1 cup)
- gray and brown food coloring
- light string (approximately 1 meter)

Procedure

The bird is fed dyed birdseed for a week so that it can become familiar with this food source. It is given no food for the six hours preceding the actual experiment. A chessboard is placed in the center of a large table, and the brown and gray pebbles are placed on the dark and light squares, respectively. The students then place 20 grains of seed (10 brown and 10 gray) randomly on the 16 squares of the board. During the experiment, the bird may be gently tied on one leg with light string or yarn approximately one meter in length to keep it from flying away.

The bird is allowed to eat any birdseed it can find on the board during a two-minute trial. After the trial, students determine and record the number of color-adapted seeds (brown seeds on brown squares or gray seeds on brown squares) the bird has eaten. The student should randomly place a total of 10 brown and 10 gray seeds on the board for each trial. The data obtained from a series of trials can then be used to determine the survival value of color adaptation.

Students may extend the experiment at a future time by using insect larvae of different colors (or any other bird food source) placed on an appropriate background. The results can be compared with those from the seed trials.

The tame birds will obviously be the limiting factor in this exercise, as it is unlikely that enough birds will be available for multiple student groups to perform this investigation simultaneously. Therefore, this activity might be done as a demonstration for the entire class or accomplished on successive days by different groups of students.

Where student groups perform the exercise, it would be possible to compare the results from day to day to see if the bird becomes more adept at locating food on the chessboard or shows an increased preference for a certain kind or color of seed.

Reference

This activity is based on J. Abruscato and L. Kenney (1972). An experiment with color adaptation. *The American Biology Teacher, 34*(3), 161, and is modified and reprinted with permission of the publisher.

VII. SIMULATING NATURAL SELECTION

This section of the monograph features a variety of models simulating the process of natural selection. These models each represent a synthesis of many of the important aspects of the Darwin-Wallace theory.

Several examples are provided, ranging from the complex design of Allen, *et al.,* which uses wild birds to generate actual numerical data, to the more traditional models involving beans, colored dots, and toothpicks. Each of these simulations highlights some unique feature of the Darwin-Wallace model, and instructors may choose one or more that are most appropriate for their students.

With several of these simulations appropriate for students of the same ability, it might be interesting to run a variety of labs in the same classroom with students from different groups reporting on conclusions reached after completing different simulations. Likewise, as with other activities, one simulation could be chosen for use as a class exercise, and another could be used later to assess authentically what students have learned.

SIMULATING EVOLUTION

Based on an original activity by
Robert C. Stebbins and Brockenbrough Allen

This simulation of natural selection uses dots of different colored paper scattered on various cloth backgrounds. Students act as predators to remove the paper dots that they are able to find. In a unique step, the remaining dots are collected and arranged by color into a graph paper histogram to help students visualize what has happened to the species variants. The survivors may be subjected to another bout of predation, accompanied by another histogram.

Other ideas include having the students wear different colored cellophane glasses to test the effects of color vision on predation or marking a pattern on the dots themselves to see if that affects the end result. Suggestions for outdoor trials with toothpick "caterpillars" with pipe cleaner bodies are also provided.

Evolutionary Principles Illustrated

- Natural selection
- Hardy–Weinberg equilibrium

Intended Audience

- Life science
- General biology

Introduction

In this simulation, a population of individuals of 10 different colors (punched out paper strips) is distributed over an imitation habitat of colorful, patterned fabric. Predators (humans) prey upon the population and remove 75%. The survivors reproduce asexually, producing three offspring like themselves, thus returning the population to its former size. Asexual reproduction is used for simplicity.

The process of predation and reproduction is repeated once or twice, after which most survivors blend with their surroundings, and the population is adapted to the color of its background. If 100 animals are used, it is easy to calculate percentages of surviving color types.

Obviously the demonstration greatly oversimplifies what happens in nature. However, it should provide a clearer understanding of natural selection than can be obtained from reading alone. The participants are involved personally in the dynamics of the population changes. Since they themselves are the predators, they can appreciate more fully the nature of the changes that take place.

Basic factors involved in natural selection are encompassed by the demonstration, even though only asexual reproduction is employed. Asexual reproduction must have preceded sexual reproduction in the evolution of life on Earth, and many organisms now living reproduce in this way. Some higher vertebrates — certain species of fish, amphibians and lizards — reproduce by parthenogenesis.

Materials (for each student group)

- 1 quarter-inch paper punch, preferably with a compartment to hold punched-out chips
- construction paper, including different shades of the same color (10 to 20 colors including black, gray, brown, and white)
- 2 (or more) pieces of fabric (3x6 feet) each of various designs and differing in basic colors
- 1 clear plastic vial or other transparent container with lid (to hold the chips)
- cellophane tape
- graph paper (four squares/inch)
- 1 black waterproof felt pen
- 3 small bowls

Procedure

Punch out 500 paper chips, 50 each of 10 different colors. Use a wide variety of colors such as red, orange, yellow, green, blue, violet, brown, gray, black and white. To speed preparation, fold the paper to four thicknesses and punch out four chips at a time. Put chips of each color in separate plastic vials and shake well to prevent clumping.

Choose fabric patterns that simulate natural environments, such as floral, leaf or fruit prints. The patterns should be of varied colors and intricate design. Test colored chips to be used against the patterns to make sure that at least some of them blend in and are hidden.

Select several designs, each with a different predominant color. It will then be possible to demonstrate the evolution of different adaptive color types from the same kind of starting population. To do so, conduct several demonstrations simultaneously and compare the surviving populations.

Conducting the Basic Simulation

Remove 10 chips from each of the 10 vials and create a population of 100 animals of 10 differ-ent colors. Assign participants to the care and handling of chips. If there are five persons, for example, each one might be responsible for counting the replicating two colors. It is important to double-check all counts at this time and on all later occasions in the simulation. If this is not done, exponential growth can lead to unmanageable population sizes.

Place all chips in a single vial and mix well by shaking. Spread out the fabric habitat on a table top and dim room lights if chips appear overly conspicuous. Empty the vial in the center of the fabric and achieve a roughly uniform distribution of chips by moving them throughout the habitat with a sliding motion of the hand. Then go over the habitat, separate the chips that may be clumped, and place them in gap areas.

Participants should stand with their backs toward the habitat to prevent locating particular chips in advance. At a signal, each predator picks up one chip at a time. After each chip is grasped the predator should place it in a container (bowl) nearby. This forces the predator to simulate common predatory behavior in which attention is centered on the prey as it is killed or carried off. Chips may be taken from any part of the habitat by sliding the hands over the habitat.

In order to ensure the survival of 25% of the chips, a quota is prescribed for each predator. The quota is determined by dividing the number of predators into the total number of chips removed. In the present example, each participant would take 15. Arbitrarily adjust counts when multiples are uneven.

The 25 surviving chips are removed from the habitat and grouped according to color type. To remove the survivors, lift the two long sides of the fabric simultaneously and shake the chips into the trough to be sure all chips have been removed. Alternatively, fold one half of the fabric over on top of the other half, spread out the fabric close to the table top with chip surface down, then lift the fabric by its four corners a few inches above the table and shake to free adherent chips.

If more than 25 chips have survived, redistribute the survivors on the habitat and remove the excess by predation in a manner described above. If there are fewer than 25, make up the difference by random selection from among those chips captured. Minor variation in numbers of survivors (two or three chips) can be accepted, and the survivor count need not be corrected if selection proceeds for only a few generations.

Arrange the survivors in a horizontal row, about one–half inch apart, placing those of each color type together. Each surviving chip produces three offspring. Place the offspring in a vertical column below each print, using chips from the reserve supply punched out earlier. (Once participants are fully aware that each chip is reproducing, they can simply determine the number of offspring by multiplication.) When all survivors have reproduced, mix them and their offspring thoroughly and distribute them as before throughout the habitat. Repeat the entire process of selection one or more times to achieve a population that closely matches its surroundings.

Preparation of Graphs

Although a colorful record of population changes can be kept through photographs or colored drawings, it is desirable to graph results as they are obtained. Place the graph paper on a firm surface and line up the chips within appropriate squares. Representatives of the starting population can be placed in order of spectral colors (red to violet) in a horizontal row at the bottom of the graph. Cover all at once with a single piece of cellophane tape. Then arrange survivors, a row at a time, in columns. Cover with vertical strips of tape. To save time and chips, use X's to show frequency of survivors.

Time Requirements

To carry selection through two generations of survivors requires approximately 20 minutes, if all "props" have been prepared in advance.

Simulations of Other Phenomena

Adaptive Radiation

The simulation can be used to show how, from the same genetic stock, differently adapted groups of organisms may arise in different environments. There are many examples among living organisms. Notable are the adaptive radiations that took place among Australian marsupials and Darwin's finches of the Galapagos Islands.

Use three or more different fabrics and start a population of identical composition on each. After two or three generations, compare the populations derived in each of these habitats. If a pale fabric (representing a desert habitat) is included, adaptation to simple and complex environments can also be compared.

Selection for Two or More Characteristics Simultaneously

In natural populations, survival frequently is a matter of chance and occurs regardless of any seemingly-useful traits possessed by the individual. Often, however, survival is greatly influenced by the individual's total array of attributes. However, at any given time and place, one or a few attributes may be particularly important.

In the basic simulation, selection was based on only one characteristic — color. To make the selection model more realistic, traits in addition to color can be introduced. Selection then has an opportunity to work on two or more characteristics simultaneously.

In addition to color, pattern, size, shape and

thickness may be used. These traits can be combined in various ways. The chart in Figure 1 illustrates results obtained when each of the 10 color varieties used were of two sizes — five small and five large. On the habitat chosen, selection favored small size and purple and blue colors.

Patterned chips can be made by placing black felt pen markings on both sides of colored paper

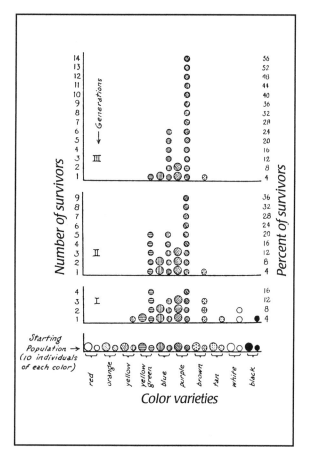

Fig. 1. Color varieties.

and punching chips from the marked strips. Half the chips of each color can be patterned and the remainder left plain. The pattern can be of ruled lines (1/8 inch apart) or closely set dots. Colors other than black can be used for patterns; felt pens come in many colors. Shape variation can be introduced by comparing half chips with round ones.

Try also thick chips. The latter can be made by gluing together two pieces of the same colored paper. Use rubber cement or hot-press photographic ready-mounted tissue. On smooth cloth backgrounds, selection often works against thick chips even though they are a precise color match, perhaps because the slightly more conspicuous shadow they cast may reveal them. This demonstrates the problem of shadow concealment that faces otherwise-camouflaged animals in nature.

Selection for a Precise Color Match to Backgrounds.

Many camouflaged animals match colors present in their habitat with remarkable precision. In this simulation, an opportunity can be afforded for the evolution of a precise color match. Included in the starting population are several shades of each of the predominant colors present in the habitat. An experiment conducted on a pale background, simulating a snow or desert habitat, may be of interest. Include in the chip population pale yellow and several shades of white (newsprint, construction paper, and so on) and make sure offspring of survivors are the proper shade.

The Founder Principle

The particular course of evolution followed by a population is greatly influenced by the hereditary composition of the original founding group. The "founder principle" is best exemplified on islands. If a rare accidental transport of a few members of a population from a mainland region to a remote island occurs, a stable colony may be established.

If the parent stock is highly variable genetically, it offers a reservoir of many different possibilities for evolutionary development on the island. The characteristics of the population that evolves there will be greatly influenced by the heredity of the particular individuals that happen to reach the new frontier. Assuming that no additional colonists arrive, it will be the heredity of these founders alone that will provide the genetic material upon which natural selection will work.

To illustrate the principle, place chips (perhaps 50, no two alike) in a container. The container represents the ancestral habitat occupied by this highly variable population. Remove 10 chips at random without looking at them. These are the colonists. Replace those removed with an identical set. Distribute the "founding" 10 individuals on a fabric representing the new environment. Increase each of the color types to 10, making a population of 100. Conduct two or three generations of selection, keeping a record of population changes as described earlier.

Remove a second random sample from the parent population on the mainland habitat and repeat the process. Compare the final population obtained in each of these experiments. Differences resulting from the accident of initial sampling will be present.

This simulation can be made more realistic (but more time consuming) by throwing a handful of chips from the parent population toward the "island" habitat, allowing them to fall short on the floor nearby. Assume that the closest 10 individuals would have reached the island and use them in starting a population. Pick up the remaining chips and return them to the container. Add replacements for those removed to restore the ancestral population for the second trial.

Investigating Predator Vision

Changes in habitat and in the variability of a population are not the only factors that can influence selection in our simulations. One can experiment with changes in the predators by providing one predator group with colored cellophane masks and another control group with clear masks. *(To make a mask, cut a rectangle of cellophane approximately 8x4 inches and attach a 12-inch strip of masking tape along one of its long sides. Tape the mask to the forehead. Red is an effective color.)* The masked predators will experience great restriction in color vision and will see the world presumably as do certain animals — in varying tones of a single hue.

Greatly limited color vision appears to be quite common among animals. It is thought to occur in such well-known predators as wolves, foxes, dogs, lions, tigers and domestic cats. On the other hand, there are animals, such as lizards, many fish and birds, that have good color vision. The subject of animal color vision, however, has not been studied sufficiently to provide conclusive answers.

Conduct two natural selection simulations simultaneously using the same starting populations on separate backgrounds but of the same pattern. In one habitat you might use predators with restricted color vision (red cellophane masks) and in the other those with normal vision (clear cellophane masks). Alternatively, use one group of predators and do the simulation twice, using first control and then red masks. Compare the populations evolved after two generations. This demonstration often yields surprising results.

Possibilities for Further Simulations

We wish to stress the open-ended nature of the basic simulation and the collateral activities that have grown out of it. Actually these simulations are experiments with variables that can be manipulated. Students should be invited to innovate and explore new avenues to understanding. Once the basic natural selection demonstration has been experienced, many people find ways to use the physical materials of the demonstration to illustrate other population phenomena.

Investigating Mutations

To simulate mutations, add a number of new chips to an "adapted" population and continue the selection process. In most cases, it can be expected that all the mutants will quickly die out, simulating what happens in nature. Occasionally, however, one can expect that one or more mutants will take hold and expand in the population.

To obtain a "take" in the short time usually available for conducting simulations, a high mutation rate may be required. It may also be necessary to introduce each mutant in sets of three or four chips. In order that the viability of the several mutants can be compared, each set must be composed of the same number of chips. Often in classroom trials, several selection simulations are conducted simultaneously on different backgrounds.

It would be interesting to introduce the same kind and number of mutants into each of the adapted populations at the same stage of their evolution (for example, after the second generation). By introducing these mutants into several populations simultaneously, the chances of mutants becoming established would be greatly increased.

One could also experiment with changing the habitat of an adapted population and noting the contribution that mutants might make to adaptation in the new surroundings. Experiments with habitat change would be desirable whether mutants are introduced or not. For example, a population evolved to match a "desert" background may be transferred to "jungle" and subjected to selection in the new environment.

Investigating Sexual Reproduction

Trials indicate that the model is capable of demonstrating dihybrid and trihybrid crosses and that it shows promise of demonstrating such phenomena as Hardy–Weinberg equilibrium and genetic drift. Special laminated chips may be used to represent homozygous and heterozygous individuals.

Heterozygote chips contain a colored inner layer that serves to code for hidden genotypic information. Special dice allow determination of offspring genotypes. The dice generate a precise simulation of the probabilities and ratios of chip genotypes resulting from any given cross between individuals. If natural selection in a sexually-reproducing population can be demonstrated, the pedagogic value of the simulations will be greatly increased.

Presently, these simulations do not show the great importance of sexual reproduction in providing the genetic variation so essential in the selection process. A sexual version could reveal how hidden variability (heterozygosity) can be made available to natural selection through the process of genetic recombination.

Investigating Selection in Predators

The effects of foraging success on the size of the population of predators can be investigated. Reproduction in predators can be geared to capture of prey. For example, a predator is required to capture a specified number of prey for the production of each offspring. If it fails to reproduce in a prescribed time, it is eliminated. In such simulations each predator feeds as rapidly as possible until a monitor calls a halt. Successful hunters increase in number. Success fluctuates with ease in detection of prey.

Each predator can be represented by different colored chips kept together at one side of the habitat. An event selector (dial with spinning arrow) can provide for random genetic changes and environmental factors affecting both predator and prey populations. For example, a predator may inherit a condition or have an accident that causes a change in vision and requires the predator to wear a red cellophane mask or search for prey with one eye covered.

Discussion

In designing this simulation of natural selection, an effort has been made to provide students with a greater understanding of the relationship between an organism's characteristics and its environment and how adaptive changes can take place in natural populations.

Questions arise as to the source of the variability present in the starting populations. The role of mutations is discussed. They are the only source of new genetic information in our asexually reproducing populations. It is pointed out that mutations which are disadvantageous under one set of environmental conditions may be advantageous under another and that traits selected in one environment may be selected against in another.

Do the terms "superior" or "inferior" (good or bad) in reference to a characteristic have any meaning if no environmental or situation context is given? Might this also be true of things other than animals (i.e., cars, books)?

What is represented by the several kinds of colored chips in the starting population? Are they varieties within one variable species or are they separate species? We have deliberately avoided classifying them. How the color types are viewed does not affect the demonstration of the natural selection principle. However, if the population is considered to be a variable species, the color variants present (if genetic) must be viewed as having arisen solely by mutation.

It is important to make clear to students the shortcomings of the present basic simulation. In particular, the lack of the great contribution to variability made by sexual reproduction and the rapidity of the simulation generation time should be noted. Students should realize that in humans and other complex organisms with a slow generation rate, vast periods of time, measured in hundreds or even thousands of years, have been involved in the processes we have simulated quickly.

This simulation of natural selection is more comparable in its rate to that of bacteria and some fungi. Furthermore, predation is merely one of many factors in natural selection. To broaden the conceptual base of the simulation, one might view the removal of chips as the decimating effect of disease, moisture or temperature extremes, or environmental contaminants. Chip color variation should then be thought of as the range in tolerance for the factor in question.

Author Acknowledgment

We have had helpful comments from the following persons who have read the manuscript and/or participated in the testing of activities presented here: Richard Eakin, Ned Johnson, Lawrence F. Lowery, John David Miller and David B. Wake.

Reference

These activities are modified from an exercise by R.C. Stebbins and B. Allen (1975). Simulating evolution. *The American Biology Teacher, 37*(4), 206–211, and are modified and reprinted with permission of the publisher.

SIMULATING NATURAL SELECTION

Based on an original activity by
R. Patterson, T. Custer, and B.H. Brattstrom

This activity presents a model for natural selection simulation including color-matching by prey (crypsis), morphologic adaptation to habit (beak length versus prey size), flock versus individual feeding success, and the concept of carrying capacity of the environment.

Evolutionary Principles Illustrated

* Selection advantage
* Adaptation
* Carrying capacity

Intended Audience

* Life science
* General biology

Materials (for each student group)

* 100 green toothpicks
* 100 red toothpicks
* tongs or tweezers (enough for half of the students in the group)
* 100 wooden matches (long-stemmed)
* 100 wooden matches (short-stemmed)
* pictures of various Galapagos finches

Procedure

Concealing Coloration

The class is divided into two equal groups. Each group includes a person to act as recorder. The group is taken to a preselected habitat, which consists of two large lawns or weedy fields. Each group walks onto one of the lawns or weedy fields. The students close their eyes while 100 red and 100 green toothpicks (cocktail or food-color-dyed) are scattered at random in each of the habitats. (This can be done before class.) The toothpicks represent insect prey.

The students then open their eyes. Pretending to be birds, they collect as many of the toothpick prey as possible in a 30-second trial. (Size of plots and amount of time spent feeding can be varied with interesting results.) They close their eyes again, and the recorder tallies and collects the toothpicks from each person. The trials are repeated until each group has collected most of the 200 (toothpick) prey.

The red toothpicks are very obvious and are picked up rapidly, but eight or more trials may be required to find all of the green toothpicks. This agrees with and supports the findings of Kettlewell (1959) on the selective advantage of habitat-matching in moths.

The class is next taken to a brown dirt habitat, where 200 red and green toothpicks are distributed and the hunt for them is repeated. Both colors of prey (toothpicks) are easily seen here and are selected against by the birds (students). Usually fewer than four trials are needed to collect most of the toothpicks.

Morphologic Adaptation

To show morphologic adaptation within a habitat, beak length in birds as seen in the Galapagos Island finches (Lack 1953) can be easily investigated. Two groups of students are taken to the grassy or weedy plots as before. Half of each group is provided with kitchen tongs or long forceps, and these students act as the long-beaked birds. The other half of each group is asked to pick up prey with only one hand, acting as the short-beaked birds.

Wooden matches are scattered over the two plots — long-stemmed matches on one plot and short-stemmed matches on the other. In either case, the matches are to be picked up with the hand or tongs and transferred to the other hand for holding. The data show the disadvantage of using tongs (long beaks) to pick up the short-stemmed matches (small prey) but not the long-stemmed matches (larger prey).

Feeding Efficiency

Flock versus individual feeding efficiency can be studied by using 200 green matches on the grassy plots. Half the class "feeds" as individuals; the other half feeds as a flock or herd, each student remaining within 30 cm of his/her neighbor. The flock or herd usually will collect more food items because of its greater efficiency in finding prey in a restricted locality, its cooperative strategy in hunting, and its social facilitation, as was noted by Etkin (1967) in both mammals and birds.

The effect of injury on feeding efficiency can be studied by using 100 red toothpicks on the green plot. In one group, the students cover one eye with a hand, simulating an eye injury; in the other group, the students simply place their unused hand on top of their head. The data usually show higher efficiency in the binocular birds than in the injured, monocular birds. The role of injury and illness in decreasing the chances of survival is well known for both individuals and groups of animals; for example, Washburn and DeVore (1961) noted this in baboon troops.

Carrying Capacity

Carrying capacity can be illustrated on two grassy plots of equal size. Scatter 200 toothpicks on each plot. Start with only six students (birds) to a plot. After each trial, while the number of prey items captured is being tallied, cast 26 new toothpicks onto the plot selected for prey growth. The birds on the other plot, which has no prey growth, are soon observed to "starve," but the competition for prey becomes rigorous on the plot with prey growth.

The model can be strengthened and altered by changing the numbers of birds on each plot. For example, have students consider what would happen with longspur, a tundra bird that requires three prey items every 15 minutes for eight hours every day, on average (Custer 1971).

Discussion

Figure 1 on the next page is a sample data sheet to be used by one of the recorders. Data can be accumulated in the field and graphed later in the laboratory. A graph could show the cumulative number of prey removed from the habitat under a given condition of predation (Figure 2), or it could show the number of prey removed under two conditions: the latter would show the number of prey remaining in the habitat per trial, and the former would show the number of prey still to be removed from the habitat.

The students can also tally the data for each individual and then rank the birds as to their efficiency in each habitat. Usually students will discover that a bird is more efficient in one habitat than in another. Data on sex can also be tallied; on average, males will have collected more prey items than the females.

Condition:	Bird No.	Trials and Prey Type												
		1		2		3		4		5		6		7	
Grass		R	G	R	G	R	G	R	G	R	G	R	G	R	G
	1	12	14	7	8	7	3	1	1	0	2	0	0		
	2	9	4	8	5	5	4	0	2	0	1	0	1		
	3	5	2	6	5	3	1	3	1	0	1	0	1		
	4	4	4	3	4	2	3	0	2	0	0	0	0		
	5	4	5	5	3	1	2	0	0	0	0	0	0		
	6	7	6	3	6	1	4	4	4	0	1	0	0		
	Sum	41	35	32	31	19	17	8	10	0	5	0	2		
	Cum f	41	35	73	66	92	83	100	98	—	98	—	100		

Fig. 1. Data sheet (abbreviated) used by the recorder.

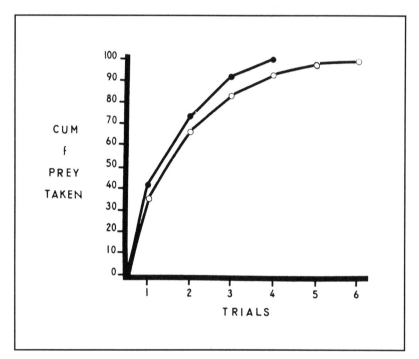

Fig. 2. Capture of red and green prey in a green grass habitat.

Reference

This activity is based on an original exercise by R. Patterson, T. Custer and B.H. Brattstrom (1972). Simulations of natural selection. *The American Biology Teacher, 34*(2), 95–97, and is modified and reprinted with permission of the publisher.

DEMONSTRATING NATURAL SELECTION THROUGH THE SURVIVAL VALUE OF CRYPTIC COLORATION AND APOSTATIC SELECTION

Based on original activities by
J.A. Allen with K.P. Anderson, S.R. Ashbourne, J.M. Cooper, and G.M. Tucker

This exercise simulates the effect of natural selection by feeding birds on both conspicuous and prey items that blend in with the background (crypsis) and demonstrates the maintenance of color polymorphism when predators select common forms of prey while ignoring rare ones (apostatic selection).

Evolutionary Principles Illustrated

- Adaptations
- Apostatic selection
- Cryptic coloration

Introduction

The activities presented here have been extracted from three different sources; however, all use the same basic method. Artificially produced baits made of lard and flour dyed with food coloring are counted and scattered on some background. (This background may be the ground itself or the surface of a feeding table.) These baits are then eaten or ignored by wild birds. The number and diversity of birds visiting the area will depend on where it is situated, but this activity has been done successfully with blackbirds *(Turdus merula)*, songthrushes *(T. philomelos)*, robins *(Erithacus rubecula)*, starlings *(Sturnus vulgaris)*, and house sparrows *(Passer domestics)*, among others.

It is sensible to pretrain the birds for several weeks to visit the area by feeding them bread, wild bird seed, etc., each day for at least one week before the start of the experiment. To encourage them to appear during the experiment proper, the pretraining food should be put out at roughly the same time of day as the baits would be.

Artificial "prey" items are useful because they can be standardized for color, size and shape and can be produced in large numbers. Furthermore, their properties can be modified to match those of the background.

There is also justification in using a background whose color can be altered because the option exists of attaining the color match by adjusting the color of the prey, the background, or both. Moreover, by repeating the experiment on more than one background color, it is possible to control for color preferences caused by factors unrelated to crypsis.

We believe the methods described here could be used by students of various age groups (perhaps with the omission of statistical tests). Whatever the age of the participants, the results from the simulation should encourage classroom discussion on the role of natural selection in the evolution of real organisms.

Intended Audience

- Advanced biology

Materials (for each student group)

- lard (340 g)
- plain flour (666 g)
- orange and green food coloring
- caramel dye
- white and brown stones (approximately 10 mms), perhaps from commercially-available washed river gravel
- plywood and molding
- ground stakes (20)
- string (1 ball)

Procedure

Preparation of the Baits

The baits are made from lard and plain flour, in a 1:3 ratio by weight, and edible food coloring in the following technique suggested by Turner (1961) and Allen and Clarke (1968). After thorough blending with a sturdy food mixer or by hand, produce brown pastry by mixing in 4 cc of orange dye and 6 cc of caramel dye to every 1000 g of dough. Other colors of pastry bait may be made by mixing in sufficient food dye to match (by eye) one of the background colors of interest. White pastry has 10 cc of distilled water added to every 1000 g of dough. To the human eye, the two types of bait appeared equally cryptic when resting on their respective matching backgrounds.

A modified mincer attachment on the food mixer may be used to produce long "worms" of pastry about 6 mm in diameter. Alternatively, the dough can be forced through holes provided with a Playdoh Fun Factory® (Cooper, 1984). The extruded worms of colored dough are then cut into cylindrical baits (length 7 mm, diameter 6 mm) and stored in plastic boxes in a refrigerator until required.

Preparation of the Feeding Station

The basic background is a square wooden tray

(410 mm x 410 mm) filled with 2500 g of small stones of a single color; there is a choice of colors to make the prey either cryptic or conspicuous. The tray consists of a base of flat plywood with a strip of wooden beading nailed along each side to give a 10-mm vertical lip; for drainage in wet weather, it has four 5-mm holes drilled in the base. It is designed to fit on a bird table made from a square piece of plywood the same size as the tray and screwed through its center into a 2-M vertical wooden pole dug into the ground.

An Introduction to Cryptic Coloration

An organism is said to be "cryptic" when it matches the coloration of its immediate background (Edmunds 1974; Endler 1981; Allen & Cooper 1985). The adaptive significance of this color match is easy to comprehend; in prey, it decreases the chances of detection by predators, and in predators, it decreases the chances of detection by prey. Although work on free-living and captive predators has produced abundant evidence for the selective advantage of crypsis in prey, most of these experiments are not practicable for use in school. This laboratory activity demonstrates that crypsis has survival value.

One of the clearest examples of the power of natural selection is implicit in the match that many palatable prey animals have with the coloration of the background. The more perfect this "crypsis" (Edmunds 1974; Allen & Cooper 1985), the greater the chances of avoiding detection by predators dependent on sight. No great stretch of the imagination is needed to understand why an uneaten individual is more fit than an eaten one, and it is hardly surprising that biology teachers worldwide continue to cite the story of selective predation in the peppered moth *(Biston betularia)* population (Kettlewell 1973) as a classic example of natural selection.

Kettlewell (1955, 1956) tested the hypothesis that selection by sight-dependent predators was responsible for the rise in the frequency of the melanic (dark-colored) forms of *Biston betularia* following the Industrial Revolution in England. He performed a series of experiments to check whether melanics were at an advantage over the lighter-colored "typicals" because their coloration was a better match to the moths' soot-covered, daytime resting sites.

He presented caged and free-ranging wild birds with equal numbers of the two forms against dark backgrounds (which made the typical cryptic). The birds removed a higher proportion of conspicuous moths from each background than was expected by chance.

Another classic example of the efficacy of selective predation is provided by "habitat correlation" in *Cepaea nemoralis* and *C. hortensis,* two species of land snail that are highly polymorphic for the coloration of their shells (Cain 1983a & 1983b). In some populations, those varieties (morphs or forms) that, to the human eye, are the most cryptic also tend to be unbanded shells in woodland and yellow five-banded shells in grassland (Cain & Sheppard 1954). All the morphs — and there are many — are inherited, and they occur in proportions higher than would be expected from recurrent mutation alone.

By carefully monitoring predation by songthrushes (*Turdus philomelos*) on two isolated populations of *Cepaea nemoralis,* Sheppard (1951) was able to show that cryptic morphs are at a selective advantage because they are the ones the birds are most likely to overlook.

Birds are undoubtedly one of the most significant groups of terrestrial predators that hunt by sight. They remove moths, snails, and the more obvious forms of praying mantis (di Cesnola

1904) and frogs (Tordoff 1980). In this activity we present simple methods to demonstrate the general point that wild birds tend to choose conspicuous varieties of prey and overlook cryptic ones. Bantock and Harvey (1974) provide a useful review of a variety of methods that can be used to simulate selective predation by birds and humans.

Procedures for Demonstrating and Studying Cryptic Coloration

The method depends on using white and brown pastry "baits" as the prey, white and brown stones as the backgrounds, and wild birds as the predators.

The 10 mm diameter stones that will serve as the substrate are washed river gravel of the type used for surfacing driveways. White and brown stones were produced by sorting the gravel into the two color categories. The container for the stones can be a plywood tray with a lip around the sides and divided by a piece of wood into two rectangles of equal size. About 100 g brown gravel (approximately 2700 stones) is poured into one half and the same quantity of white gravel into the other half. The tray is placed on a sturdy table away from pedestrian traffic in an area where birds have come to feed.

In Experiment 1, the birds are given a choice between equal numbers of the two colors strewn randomly over each of the two backgrounds. The null hypothesis is that equal numbers of the two colors should be taken from each of the two backgrounds.

For 15 nonconsecutive days, scatter 10 white baits and then 10 brown baits randomly over each of the two backgrounds. Predation from a given background is recorded when approximately 10 baits have been removed (which was not easy to judge) or, failing this, at the end of the day. Search the trays thoroughly for uneaten prey. (This is extremely important because it is essential to eliminate the possibility that humans, not birds, were responsible for any selection which was subsequently detected.)

Table 1 gives the results from Example Experiment 1. Near the foot of the table, the grand totals of baits removed from each background are compared with the numbers expected, assuming no selection. It is clear that an excess of browns was taken from the white background and an excess of whites was taken from the brown background. Each of these deviations from a ratio of 1:1 is statistically significant

sign–ranks test (Siegel 1956). There is an even greater tendency for the conspicuous color to be taken on the brown stones.

In the second example experiment, 20 white baits were scattered on each of the two backgrounds for 10 trials. Our new null hypothesis was that there should have been no statistically significant difference in the number of baits removed from the two backgrounds. The procedure was then repeated with brown baits. For each trial we had intended to count the numbers of baits eaten from the two backgrounds when

Trial number	White background Numbers eaten		Brown background Numbers eaten	
	white	brown	white	brown
1	4	3	7	0
2	1	0	2	0
3	2	4	4	3
4	6	3	9	4
5	4	6	6	3
6	5	7	9	6
7	3	6	7	1
8	3	6	4	3
9	2	6	7	3
10	0	6	8	1
11	3	8	6	2
12	0	3	9	7
13	3	7	9	4
14	2	8	8	5
15	1	6	9	3
Grand totals	39	79	104	45
Expected 1:1	(59)	(59)	(74.5)	(74.5)
	$\chi^2(1:1) = 13.56$, 1 d.f., p < 0.001		$\chi^2(1:1) = 23.36$, 1 d.f., p < 0.001	

Table 1. Numbers of baits eaten daily in Experiment 1.

when tested by chi–squared (bottom row, Table 1).

Another way of analyzing the data is to examine the selection within the individual trials. Of the 15 trials on white stones, 12 deviated from 1:1 in the direction predicted, and when the magnitude of the deviations is also taken into account, this trend is found to be statistically significant — T = 10.5, p < 0.01, Wilcoxon matched–pairs

20 baits in total had been removed.

Table 2 gives the results from Example Experiment 2. The grand totals depart from the expected 1:1 ratios in the directions predicted by the hypothesis that the birds found white baits easier to detect on brown stones and brown baits easier to detect on white stones, although this deviation is statistically significant for the brown prey only.

Examination of the individual trials confirms that the baits were more likely to be eaten when resting on the nonmatching backgrounds. In seven of the trials with white baits, a greater number of baits were removed from the brown background in eight of the 10 trials (T = 4.5 p<0.02).

An Introduction to the Maintenance of Color Polymorphism by Apostatic Selection

Color-pattern polymorphisms are widespread in most animal phyla and many groups of plants,

One idea that has become popular in some quarters is that foraging predators might concentrate on common morphs and ignore rare ones (Clarke 1962; Moment 1962; Greenwood 1984; Allen 1988). As a result of this "apostatic selection" (Clarke 1962), or "switching" (Murdoch 1969), fitness would be inversely related to morph frequency, and thus polymorphism would be actively maintained (Figure 1).

Several mechanisms could cause the behavior, of which the acquisition of "search images" (Dawkins 1971) for common prey is but one

| | White prey | | | Brown prey | |
| | Numbers eaten from | | | Numbers eaten from | |
Trial number	white background	brown background	Trial number	white background	brown background
1	16	18	11	13	8
2	19	17	12	15	11
3	15	19	13	14	13
4	16	15	14	11	12
5	14	18	15	13	5
6	14	17	16	16	9
7	9	16	17	16	11
8	11	15	18	9	11
9	12	18	19	17	12
10	16	16	20	18	12
Grand totals	142	169		142	104
Expected 1:1	(155.5)	(155.5)		(123)	(123)

$$\chi^2(1:1) = 2.34, \text{1 d.f., not significant} \qquad \chi^2(1:1) = 5.87, \text{1 d.f., } p < 0.05$$

Table 2. Numbers of baits eaten daily in Experiment 2.

and there is evidence from several of them that the variation has existed for thousands of years. A variety of agents of selection have been identified — for example, in the snail *Cepaea,* climate and predations are undoubtedly important (Jones et al. 1977; Clarke et al. 1978; Cain 1983a, 1983b). However, most of these agents act directionally, removing certain morphs from the population while favoring others. How, then are the populations kept variable? There may be no simple answer.

(Murdoch et al. 1975; Greenwood 1984, 1985). Predators that concentrate on common prey may benefit by optimizing their rate of food intake (Hubbard et al. 1982).

Procedures for Demonstrating and Studying Apostatic Selection

Apostatic selection, like cryptic selection, can be tested by feeding artificial prey to birds. If the aim is simply to measure selection on the prey

A_1=number of prey 1 available
A_2=number of prey 2 available
e_1=number of prey 1 eaten
e_2=number of prey 2 eaten

Thus $A_1/(A_1+A_2)$ is the proportion of prey 1 available to a predator and $e_1/(e_1+e_2)$ is the proportion of prey 1 eaten.

A—Apostatic selection occurs when the proportion of a prey type eaten by a predator (solid line) deviates from the relationship expected in the absence of selection (broken line) such that, when common, a higher proportion than expected are eaten and, when rare, a lower proportion than expected are eaten. In this example there is no additional frequency-independent selection.

B—Apostatic selection is less easy to detect when there is frequency-independent selection against one of the prey types: in this example prey 1 is nearly always eaten in a higher proportion than expected by chance.

C—Apostatic selection detected by fitting the model of Manly (1985), where, either:

$$\beta = \frac{\log_{10}\left(\frac{A_1}{A_1 - e_1}\right)}{\log_{10}\left(\frac{A_1}{A_1 - e_1}\right) + \log_{10}\left(\frac{A_2}{A_2 - e_2}\right)}$$

(for experiments where prey are notreplaced until the end of the trial),

or:

$$\beta = \frac{e_1/A_1}{e_1/A_1 + e_2/A_2}$$

(for experiments wher the proportions offered have been constant by frequent replacement of eaten prey).

ß ranges from 0 to 1.0; ß=0.5 indicates no selection, ß>0.5 indicates selection against prey 1, and ß<0.5 indicates selection against prey 2. The broken line therefore depicts the relationship expected in the absence of selection. The positive slope of the solid line indicates apostatic selection. The expected equilibrium frequency is given by the intercept of the two lines; in this example there is a frequency-independent component of selection against prey 1. A negative slope would indicated anti-apostatic selection.

D—Apostatic selection detected by fitting the model of Greenwood and Elton (1979):

$$\frac{e_1}{e_2} = \left(\frac{VA_1}{A_2}\right)^b$$

where b and V are measures of the degree of frequency-dependent and frequency-independent selection respectively.

If a range of prey frequencies are used then the relationship expected in the absence of selection is indicated by the diagonal broken line in the graph. Apostatic selection is indicated when the slope (b) of the solid line is greater than unity (as here); anti-apostatic selection would be indicated when the slope is less than unity.

Fig. 1. Detection and measurement of apostatic selection when two types of prey are presented.

populations, then it is often more practicable to visit the prey at intervals and record the pooled population.

In this example experiment, 180 green and 20 brown baits are scattered randomly on a grass lawn (Allen & Clarke 1968). Predation by wild birds is recorded, and the 9:1 ratio kept constant by frequent replacement. After a week, the frequencies are altered so that browns are now nine times more common than greens.

Thirteen experiments of this design have confirmed that the birds tend to remove the common color (Allen 1976). Although the birds usually ate an excess of one of the colors, whether or not the color was common or rare, in every case the selection against the color was greater when it was common, as predicted by the hypothesis of apostatic selection.

Specific Instructions for the Apostatic Selection Activity

1. Select a site known to be frequented by ground feeding birds (grass lawns are particularly convenient.

2. Make a sufficient number of artificial prey in two colors (use green instead of white if selecting a grass lawn).

3. Decide on the prey densities and population size [i.e., 2m–2 and 200 (Allen & Clarke 1968)].

4. Use pegs to map out a grid of meter-squares to contain the prey.

5. To control for selection independent of frequency, it is important that at least two frequencies are presented, say 0.1 and 0.9 of a given color. Decide (randomly) which frequency will be presented first. Draw the grid on paper and plot a random distribution of the appropriate numbers of the two colors of prey.

6. Using the plan as a guide, distribute the prey within the actual grid.

7. Either (a) watch the grid for as long as the prey are presented (if interested in selection by known individuals or species) or (b) check the grid at frequent intervals. Record the number eaten and replace prey to maintain the 9:1 ratio. Repeat for a number of days (i.e., five days), changing the distribution at least once a day. Alternatively, count the proportions eaten after roughly a certain fraction have been eaten (i.e., 30%), replenish and repeat.

8. Repeat Steps 2-4, but with the second prey frequency.

9. Calculate a coefficient (ß) for selection against one of the prey types, using an appropriate formula depending on whether or not eaten prey were replaced during the experiment on the opposite page (Figure 1).

- For experiments with replacement, if apostatic selection has been acting, ß should be higher when measured for the population in which Type 1 was common.

- If there was no replacement, then ß can be calculated for each trial, and the statistical significance of the difference between the mean ß value for each population can be tested by a t-test or one-way analysis of variance — preferably after normalizing the data using the arcsine transformation (Sokal & Rohlf 1981).

Additional Suggestions

Little is known abut the influence of prey palatability. Monomorphism is predicted to evolve if the prey are unpalatable, because predators are more likely to learn to avoid commonly encountered morphs than they are rare ones. Pastry baits can easily be adulterated with nasty-tasting substances such as quinine salts, but the data from predation by wild birds are contradictory (Greenwood et al. 1981). One might try varying the prey density and/or form for another variable.

Author Acknowledgment

We thank the University of Southhampton for facilities, finances, and a supply of cooperative students; B. Lockyer for photographic assistance; and K.P. Anderson for his bait-making skills.

References

These exercises are extensively adapted from the following original activities. All activities are modified and reprinted with permission of the publishers.

- J.A. Allen, K.P. Anderson and G.M. Tucker (1987). More than meets the eye — A simulation of natural selection. *Journal of Biological Education, 21*(4), 301–305.

- J.A. Allen and J.M. Cooper (1988). Experimenting with apostatic selection. *Journal of Biological Education, 22*(4), 255–262.

- J.A. Allen and S.R. Ashbourne (1988). Demonstrating the survival value of cryptic coloration. *School Science Review, 69*(248), 503–505.

DEMONSTRATING THE EFFECTS OF SELECTION

Based on an original activity by
Jamie E. Thomerson

Population genetic exercises involving living organisms are difficult to design for completion within one lab period. Therefore, in the population genetics experiment described here, beans are used to represent genes in the population gene pool.

Evolutionary Principles Illustrated

• Population genetics
• Selection

Intended Audience

• General biology
• Advanced biology

Introduction

The experiment requires approximately 50 minutes to complete, is inexpensive, requires no special facilities, generates participation, and introduces students to the idea of predictable change in gene frequency as a result of selection.

The students should have had some introduction to population genetics concepts — perhaps a general treatment of the Hardy-Weinberg Law — and some explanation of the concept of selection before they attempt this exercise.

Materials (to be shared by all class members)

• pinto beans (1 lb)
• red beans (1 lb)
• coffee can

Procedure

Pinto beans are used to represent the dominant gene (R) of a pair of alleles and red beans to represent the recessive allele (r). Other items, such as beads, marbles, or corn grains, could be used instead of beans, but the two different alleles should not be identifiable by touch. In introducing the experiment, the instructor should explain that the experiment simulates a situation where individuals homozygous for a given recessive allele die before they are able to reproduce.

The original gene pool is established by pouring a pound each of red and pinto beans into a coffee can and thoroughly mixing them. This gives an original gene pool with about equal numbers of both alleles. Students in the lab are divided into 10 groups, and the coffee can is passed from group to group. One member of each group, with eyes closed, picks out at random 10 pairs of beans to represent 10 diploid individuals. The first coffee can is returned to the instructor and set aside.

Each student records the genotype of his group's 10 individuals (10 pairs of genes) on a tally sheet (Figure 1) and then reports the results to the instructor. The instructor tallies the results for the whole class and computes the gene frequency for the whole population of 100

Generation	Your Group #(7)			Class Total				
	RR	Rr	rr	RR	Rr	rr	%R	%r
O.F.	5	5	0	21	61	18	51.5	48.5
1	1	6	3	40	46	14	63	37
2	5	2	3	52	42	6	73	27
3	10	0	0	58	38	4	77	23
4	7	0	3	68	24	8	80	20
5	9	1	0	80	14	6	81	19

Fig. 1. Sample tally sheet completed by student. Each student records his/her group's results and the pooled class results.

individuals (Figure 2). These are recorded by the class as the *original frequencies* (O.F.) [Figure 1].

The instructor has each group set aside all the individuals that are homozygous recessive (two red beans) and return the rest of the genes to a second empty coffee can. (A remark to the class about not dropping any of the genes is appropriate at this time.) The instructor explains that the homozygous recessive individuals have been selected against— that they have been removed from the breeding population. The new gene frequency after the removal of the homozygous recessive individuals is then computed by the instructor. The new frequencies are recorded by the class as the first generation (Figure 1).

The instructor then adds enough genes to the second can to bring the number of genes in the gene pool back up to 200 (Figure 2). Dominant

Group	RR	Rr	rr
1	5	3	2
2	1	6	3
3	1	7	2
4	0	10	0
5	3	2	5
6	0	9	1
7	5	5	0
8	5	4	01
9	1	9	0
10	0	6	4
Total	21	61	18

$$RR \cdot 21 \times 2 \cdot 42R \Big\} \cdot 103$$
$$Rr \cdot \quad \underline{61R}$$
$$\underline{61r}$$
$$164$$

$$\%R \cdot \frac{103}{164} \cdot 62.8\%$$

$$\%r \cdot \frac{61}{164} \cdot 37.2\%$$

add 18 x 2·36 beans
.63x36·23± Pinto (R)
.37x36·13+ red (r)

Fig. 2. Instructor's tally for the class results for Generation 1. The instructor would add red and pinto beans to make 200 for the next generation.

and recessive genes are added after each round in proportion to the gene frequencies after selection in that round. This addition is neces-

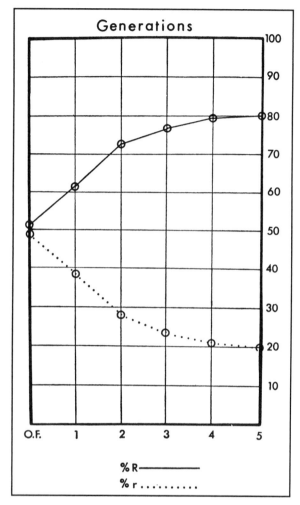

Fig. 3. Graph of the results shown in Figure 1; note flattening of curves. The graph coordinates were reproduced on the same sheet of paper as the students' tally form.

sary to the mechanics of the experiment and does not seem to confuse the students.

I originally planned to pass the second coffee can, containing the 200 genes, from group to group and let each group pull out 10 pairs at random for the next generation. In actual practice, our labs have students sitting at tables, so we have a student from each table come to the front of the room with a paper cup. The instruc-

tor divides the 200 genes among the three cups. The division is by eye, and the students quickly understand that excess genes in the cup go to those groups that are short. This does not have much effect on the randomness of drawing the genes, and the mechanics of this step can be adjusted for a wide range of situations.

As soon as each group has recorded and reported the genotypes of its 10 individuals, the student groups set aside their homozygous recessive individuals and return the rest of the genes to the second coffee can. The instructor computes the gene frequency for the population after selection; these are recorded by the students as the second generation (Figure 1). The instructor then returns the population back to 200 genes in proper proportion, and the procedure is repeated for the third generation. In one class period, five generations beyond the original population can easily be run.

Students may plot the values for each generation on the graph (Figure 3) as they are obtained, or they may wait until the end of the experiment. We have run this experiment more than 100 times and have obtained results similar to those shown in Figure 3 each time.

As would be expected, the frequency of the recessive gene drops rapidly at first, but after two or three generations, the curve flattens out. After participating in this experiment, students easily grasp the point that there is an initial rapid shift of gene frequency in response to strong selection but that deleterious recessive genes are very difficult to completely remove from the gene pool.

Even after participating in this experiment, students may not immediately appreciate the results of selection when the dominant gene is lethal. Picking 10 individual pairs out of the gene pool and then removing the individuals having the dominant genes demonstrates the point that dominant lethal genes are immediately removed from the population.

The experiment outlined previously demonstrates a very simple situation, but the procedure could easily be modified to simulate more complex situations — for example, the effects of mutation, partial lethality or selection against the heterozygote. However, the population seems to be too small to guarantee a reliable, convincing demonstration of the Hardy-Weinberg law, and doubling the size of the sample might help. Although Hardy-Weinberg is covered in detail in lecture format, there does not seem to be much interest in further laboratory investigation.

Author Acknowledgment

I first became aware of this kind of simulation experiment in a population–genetics course taught by E. Peter Volpe of Tulane University. Several of my present colleagues there have offered helpful suggestions.

Reference

This activity is modified from an exercise by J.E. Thomerson (1971). Demonstrating the effects of selection. *The American Biology Teacher, 33*(1), 43-45, and is reprinted here with the permission of the publisher.

A MODEL OF MICROEVOLUTION IN ACTION

Based on an original activity by
Larry A. Welch

The following activity is designed to help students understand the precepts of the Hardy-Weinberg principle and simultaneously permit observation of a model of evolution through natural selection.

Evolutionary Principles Illustrated

• Adaptation
• Hardy-Weinberg equilibrium
• Selection

Introduction

This activity uses students as predators equipped with a variety of prey-capturing structures, such as knives, forks, spoons, forceps and hands, in much the same fashion as in the related activity, *Birds and the Beaks*. The prey are ordinary dried beans of several colors. When these "prey" are distributed around the "environment," the "predators" begin capturing prey quickly.

Prior to this activity, it is important to discuss the Hardy-Weinberg principle if that aspect of this investigation is to be illustrated. Knowledge of Hardy-Weinberg equilibrium will help insure that students have the background to establish that microevolution (change in gene frequencies) is indeed occurring in the experimental populations. Hardy and Weinberg established the fact that sexual reproduction by itself will not usually result in changes in gene frequency. The Hardy-Weinberg expression is:

$$(p + q)2 = p2 + 2pq + q2$$

where:

p = the frequency of allele A
q = the frequency of allele a, and
$p + q = 1$.

Hardy and Weinberg independently arrived at the same conclusion when they established the principle that gene frequencies will not change in population if there is:

1. Absence of random mating.
2. No migration (in or out).
3. No mutation (or equal mutation).
4. No natural selection.
5. No genetic drift (random change of allele frequency as occurs in small populations).

Biologists know that all of these phenomena can and do act on populations. Therefore, an evaluation of changes in gene frequency becomes a mechanism for evaluating evolutionary direction and rate.

Intended Audience

• General biology
• Advanced biology

Materials (for each student group)

• plastic drinking cup
• capturing device for each student, such as a plastic spoon, a fork, a knife, forceps or a hand

- package of dried beans of each of the following colors: white, red (brown), spotted and black
- computers with a spread-sheet to calculate and graph results
- data sheets on which to record results

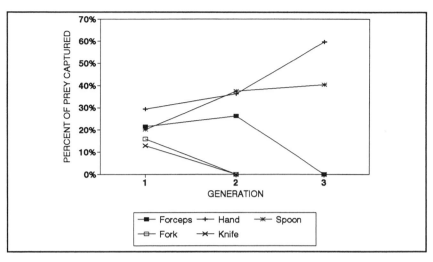

Fig. 1. Selection among predators.

Discussion

The activity works well with classes up to 24 students. Above that number, it may be necessary to modify the procedure to compensate for the large number of "predators."

The time required is approximately two hours. If your class periods are shorter, consider conducting the hunts during one class period, organizing data in a second period, and analyzing data in a third period.

Space requirements are minimal and, if class size permits, can be as small as an area 20' x 20'. Grassy lawn is the preferred surface, but almost any type of surface (grass, concrete, asphalt, etc,) will do.

Students may use a form similar to that in Table 1 (following this activity) to record the number of "kills."

Calculated cells within the table are best left to the computer. Hand calculations are good practice for the student. If used, however, your class will probably run out of time before achieving your teaching objectives.

Tables 2-4 (see page 106) are copies of spread-sheets (complete with data) that can be used to enter class records. Once entered, graphing the data is not difficult, and you will find that graphs (see Figures 1 and 2) will be appreciated by visual learners and useful to all students in depicting emerging trends.

Specific Directions

1. Count out exactly 100 dried beans of each of the four colors. Mix these together thoroughly in a single container and spread them evenly over the "habit" surface.

2. Upon an established signal, predators are permitted to begin capturing prey, but they must observe the following rules:

a) At the instructor's signal, predators are to begin hunting and continue for three minutes. During this time, the predators will attempt to capture as many prey as possible, without regard for color.

b) Predators must use their capturing devices to capture their prey.

c) Predators may not scoop prey from the ground with their cup. (The cup must not touch the ground.)

d) At the sound of the "stop" signal, the class

must stop. Prey in the capturing device but not in the cup must be released.

3. Each predator determines the number of prey captured. All predators using the same capturing device aggregate their totals, and the total number captured is entered on the computer (or data sheet).

4. The average number of prey captured for each type of capturing device is determined, and those predator types not capturing at least the mean number of prey are now "extinct." (These students return to the activity as "offspring" of those predators who captured more than the mean number of prey.)

5. From the totals of each color of bean captured, natural selection may be observed directly. That is, there will be a natural tendency for one of the colors to be more commonly captured and others to be less frequently captured.

6. From the number of beans of each color

captured, determine the number of beans of each color still remaining in the environment. The computer spreadsheet will calculate this information for you. (A complete printout of cells and formulas for the spreadsheet appears in Figure 3 on page 107.)

7. Assume that each prey specimen remaining in the environment will reproduce. Count out one bean of the appropriate environment. If 65 red beans were captured, you would know that there are 35 remaining in the environment that can reproduce. In this mode, we are ignoring other forces that tend to decrease populations. Therefore, count out 35 additional red beans to be added to the environment before the next hunt begins.

8. Repeat this procedure for each of the colors of prey. Record the new "beginning" population sizes and return the predators to the field for another three-minute hunt.

9. Repeat as many times as the class period permits and keep accurate records of changes in population numbers of both prey and predators.

Fig. 2. Selection among prey.

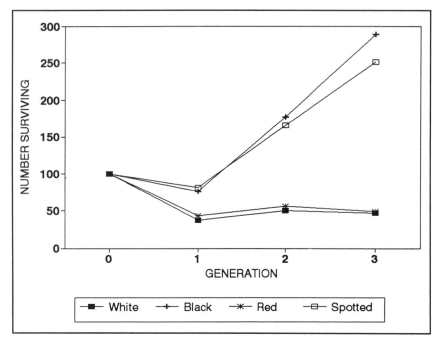

10. Divide the class into groups to analyze and report to the rest of the class what happened to each kind of predator and prey. A master data form on the chalkboard provides an opportunity for students to enter their contributions in the appropriate grid and gives the entire class access to the data.

11. You may wish to construct a "super graph" on which you plot the ascension of successful populations and the demise of unsuccessful populations.

Table 1. Sample data sheet—Number captured. Generation _____						
	White	Black	Red	Spotted	Total	Percent Captured
Forceps						
Hand						
Spoon						
Fork						
Knife						
Total Kills						
Survivors						
% Surviving						

Table 1. Sample data sheet — Number captured. Generation ___

12. Ask your students to prepare a written report of what happened in this mock predator/prey interaction. Be sure to ask them to explain their understanding of why some creatures became more numerous and others became less numerous.

Your students will begin to grasp the concept of change in populations over time and to recognize that populations, not individuals, evolve. Microevolution in action opens doors for discussion of topics in genetics, population biology, competition and natural selection. It also presents evolutionary concepts in a non-threatening fashion and stimulates discussion and interaction among students and between students and the instructor.

If possible, run the laboratory procedure for one class from a lawn environment and a second one from a snow-covered area. The differing results emphasize the significance of environment in survival of organisms and underscore the fact that each organism lives or dies based on its inherited characteristics and the environment in which it is found — microevolution in action!

Reference

This activity has been adapted from an exercise by L.A. Welch (1993). A Model of microevolution in action. *The American Biology Teacher,* 55(6), 362–365, and has been modified and reprinted with permission of the publisher.

(Top) Table 2. Predator/prey interaction — Generation 1. (Middle) Table 3. Predator/prey interaction — Generation 2. (Bottom) Table 4. Predator/prey interaction — Generation 3.

Population Prey Color	100 White	100 Black	100 Red	100 Spotted	400 Total	Percent Captured
Forceps	15	5	13	2	35	21.47%
Hand	22	9	15	2	48	29.45%
Spoon	11	2	11	9	33	20.25%
Fork	10	3	8	5	26	15.95%
Knife	5	5	10	1	21	12.88%
Total Kills	63	24	57	19	163	
Survivors	37	76	43	81	237	
% Survived	37.00%	76.00%	43.00%	81.00%	59.25%	

Population Prey Color	111 White	228 Black	129 Red	243 Spotted	711 Total	Percent Captured
Forceps	21	18	12	18	69	26.34%
Hand	22	13	31	29	95	36.26%
Spoon	18	20	30	30	98	37.40%
Fork	0	0	0	0	0	0.00%
Knife	0	0	0	0	0	0.00%
Total Kills	61	51	73	77	262	
Survivors	50	177	56	166	449	
% Survived	45.05%	77.63%	43.41%	68.31%	63.15%	

Population Prey Color	100 White	354 Black	112 Red	332 Spotted	898 Total	Percent Captured
Forceps	0	0	0	0	0	0.00%
Hand	32	42	38	44	156	59.54%
Spoon	21	23	25	37	106	40.46%
Fork	0	0	0	0	0	0.00%
Knife	0	0	0	0	0	0.00%
Total Kills	53	65	63	81	262	
Survivors	47	289	49	251	636	
% Survived	47.00%	81.64%	43.75%	75.60%	70.82%	

Fig. 3. Spreadsheet cell formulas.

A1:   [W11] 'Predator/Prey Interaction – Generation One
A3:   [W11] 'Population
B3:   [W8]   100
C3:   [W9]   100
D3:   [W8]   100
E3:   [W8]   100
F3:   [W7]   @SUM(B3 . . E3)
G3:   'Percent
A4:   [W11] 'Prey color
B4:   [W8]   'White
C4:   [W9]   'Black
D4:   [W4]   'Red
E4:   [W8]   'Spotted
F4:   [W7]   'Total
G4:   'Captured
A5:   [W11] 'Forceps
B5:   [W8]   15
C5:   [W9]   5
D5:   [W8]   13
E5:   [W8]   2
F5:   [W7]   @SUM(B5 . . E5)
G5:   (P2)   (F5/F11)
A6:   [W11] 'Hand
B6:   [W8]   22
C6:   [W9]   9
D6:   [W8]   15
E6:   [W8]   2
F6:   [W7]   @SUM(B6 . . E6)
G6:   (P2)   (F6/F11)
A7:   [W11] 'Spoon
B7:   [W8]   11
C7:   [W9]   2
D7:   [W8]   11
E7:   [W8]   9
F7:   [W7]   @SUM(B7 . . E7)

G7:   (P2)   (F7/F11)
A8:   [W11] 'Fork
B8:   [W8]   10
C8:   [W9]   3
D8:   [W8]   8
E8:   [W8]   5
F8:   [W7]   @SUM(B8 . . E8)
G8:   (P2)   (F8/F11)
A9:   [W11] 'Knife
B9:   [W8]   5
C9:   [W9]   5
D9:   [W8]   10
E9:   [W8]   1
F9:   [W7]   @SUM(B9 . . E9)
G9:   (P2)   (F9/F11)
A11:   [W11] 'Total Kills
B11:   [w8]   @SUM(B5 . . B9)
C11:   [W9]   @SUM(C5 . . C9)
D11:   [W8]   @SUM(D5 . . D9)
E11:   [W8]   @SUM(E5 . . E9)
F11:   [W7]   @SUM(F5 . . F9)
A12:   [W11] 'Survivors
B12:   [w8]   (B3-B11)
C12:   [W9]   (C3-C11)
D12:   [W8]   (D3-D11)
E12:   [W8]   (E3-E11)
F12:   [W7]   (F3-F11)
A13:   [W11] '% Survived
B13:   (P2)   [W8] (B12/B3)
C13:   (P2)   [W9] (C12/C3)
D13:   (P2)   [W8] (D12/D3)
E13:   (P2)   [W8] (E12/E3)
F13:   (PS)   [W7] (F12/F3)

VIII. PROPOSING PHYLOGENIES

One of the goals of taxonomy is to provide an outline of "descent with modification" or, in common language, to produce *family trees*. Taxonomists use a wide variety of evidence to produce natural groupings of organisms that are both related to each other and descended from or ancestors of other more distantly-related species.

Students develop the ability to classify quite early in their intellectual development but, for the most part, use superficial or unimportant characteristics in developing their personal taxonomies. This type of artificial classification explains why whales and fish are seen as close relatives even by many adults. Once students can look past superficial characteristics, such as color or basic shape, they can begin to "weight" some traits or characteristics as more important than others in their proposals of relationships and lines of descent.

The various activities in this section afford teachers wonderful case studies of classification. Students examine nuts, bolts, laboratory glassware, aluminum pull tabs from beverage containers, and imaginary creatures called "Caminalcules" to propose classification schemes. In addition, following each proposal of a "relationship" with any of these objects, students are asked to defend their choices to help them become more familiar with the notion of natural vs. artificial classification plans.

Using the method outlined above, students will become much more familiar with the actual process of classification, which at its core, is a human construct. As in other sections, a wide variety of activities is presented so that some exercises can be used for instruction and others applied to authentic assessment.

USING ALUMENONTOS TO INTRODUCE EVOLUTION AND PHYLOGENY

Based on an original activity by
Steven J. Hageman

This activity uses different types of pull tabs from aluminum beverage cans ("alumenontos") to represent organisms (taxa) that seem to have had a common ancestor. Students engage in exercises where they propose a phylogenetic relationship between these organisms. In addition, ideas showing how classification and biostratigraphy may also be demonstrated by using these curious creatures are also suggested.

Evolutionary Principles Illustrated

- Systematics
- Biostratigraphy
- Phylogeny

Introduction

Availability of adequate specimens for teaching general paleontologic principles is often a problem. This is because effective demonstration of paleontologic principles of taxonomic hierarchy, evolution, phylogeny, and biostratigraphy requires many well preserved, related specimens from a range of geologic times. Even large collections of a wide variety of fossil groups ideally suited for teaching systematics and morphology may not be appropriate.

When an adequate collection is available, students still have difficulty with subtle techniques, such as species recognition, because such concepts require considerable biological knowledge. Problems encountered with systematics can limit the use of biostratigraphic and phylogenic exercises that have recognition of discrete taxa as a prerequisite.

Therefore, objects well suited for teaching should be relatively simple, yet diverse and abundant, and students should have no preconceived ideas about their classification. The tabs used to open aluminum beverage cans fit these criteria. These "alumenontos" are treated as skeletons, rather than complete organisms, to simulate the problems of paleontologists, who work most frequently with hard parts.

The following is an outline of several of the paleontologic principles that can be introduced with alumenontos. The purpose of this activity is not to introduce the principles themselves but to show how they are reflected in an alumenonto model.

Intended Audience

- General biology
- Advanced biology

Materials (for each student group)

- 100 assorted aluminum can pull tabs

Procedure

The students are divided into small groups, and each group is given approximately 100 mixed alumenontos and told to sort them into species.

Students discover that, even with such relatively simple objects, it is possible to create two identical groups or even to further subdivide groups on closer inspection. Alumenontos have an advantage over real organisms in that the "species" (discrete types) are clear.

Students should be made aware that there is virtually no intraspecific variation among alumenontos, which is often not true of real organisms. It may even be appropriate to show real examples of extreme intraspecific variation to emphasize the point. Organisms that experience a strong environmental influence on their morphology, such as oysters, make good examples.

The students select a representative from each species (the concept of type specimens can be introduced at this time, if appropriate), and the instructor assigns numbers to each morphotype so that students can refer to taxa by numbers for later comparison. Labeled adhesive tape attached to representatives of each taxon works well. After the specimens have been sorted into species, the class characterizes the alumenontos group as a whole.

Many questions are opened for discussion, such as the proper orientation of the organism, the composition of the skeleton, symmetry elements, and whether the skeleton is internal or external. Some other questions that arise are: Is each object an organism in and of itself or simply a small element of one organism? Are the objects molts representing the ontogeny of several species? Students can speculate on functional morphology of alumenontos and the life mode of the organism from which they came (members of the phylum *Alumentophora,* of course).

Obviously, there are no right or wrong answers to these questions when applied to alumenontos. What is more important is that students discuss what constitutes valid evidence to answer these questions. Students should realize that this part

of the exercise is not unlike working with extinct taxa that have no living relatives. "How could you tell if...?" may be a more appropriate preface for many of the questions.

Students are then instructed to group their type specimens into a taxonomic hierarchy. They make a list of all the characteristics that are used to classify alumenontos: short lists of characters that typify species within each genus and separate lists of characters used to differentiate among genera. Classifications are then compared among groups of students, and the reasons for different interpretations (different characters chosen and varied degrees of weighing applied) are discussed. Once again, there are no right or wrong answers, although some classifications are more defensible than others.

A potential weakness of the alumenontos model may be its limited diversity (relative to the 37 "species" of the Burns' [1968] hardware model). However, in one possible classification of readily available material, I "found" 13 species, five genera, three families, and two classes (see Figures 1–5 on pages 115-116). This seems adequate to demonstrate the concepts. The concept of homology is introduced with the specimens shown in Figure 6 (see page 116).

Regarding the questions of sexual dimorphism versus intraspecific variation versus ontogenetic variations, subspecies are introduced with the specimens shown in Figure 7 on page 116. Once again, the questions are phrased in the context of, "What lines of evidence would you seek in order to decide whether these are sexual dimorphs of one species or two different subspecies?"

There is no great degree of size variation among alumenontos, so clear developmental sequences are not readily modeled. Unfortunately, this precludes studies designed to recognize differential growth patterns. However, speculation as to why there is an absence of size variation allows for a great deal of discussion, with many plausible answers.

The students are then instructed to construct a phylogenetic diagram representing the evolutionary sequence of alumenontos. When finished, they compare their taxonomic hierarchy to their phylogenetic reconstruction to see if the two are compatible. If not, they discuss the problems encountered and decide whether they wish to alter their classification or phylogenetic interpretation.

It is also interesting to note whether similar characters appear at different times in their phylogenetic reconstructions. In addition, the students compare phylogenies among groups of students and discuss different interpretations. It soon becomes clear that different workers perceive primitive and derived characters differently. Figure 8 shows a possible phylogenetic reconstruction for alumenontos.

Alumenontos can be used to introduce many biostratigraphic principles. For example, in one laboratory exercise, students are given copies of Table 1 (without species ranges or zone columns completed) and asked to label examples of each of the 13 species. Then, on a separate sheet, they are given the stratigraphic ranges of each species and instructed to fill in the range columns of Table 1, corresponding to the time column.

In the first part of the exercise, students are given five separate assemblages of alumenontos and are instructed to work out the time range for each of the five samples. The students are told that the samples are exhaustive (level of classification) for a given locality, so absence of a taxon is as important as presence. For example, if an assemblage consists of species numbers 2, 4, 6, 8, and 10, the students must first identify them as such and then apply biostratigraphic principles to recognize that the assemblage came from times 13-16.

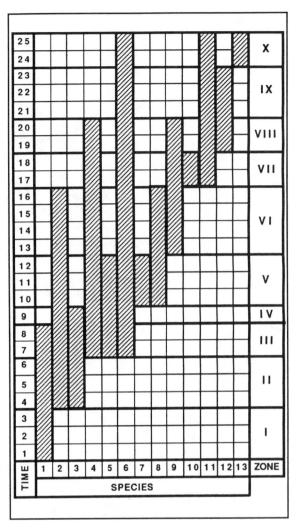

Table 1. Units of time are represented on the left and species number across the bottom. Species duration is represented by cross hatching, and a zonation scheme is shown on the right.

In the second part of the exercise, the students construct a biostratigraphic zonation scheme for the 13 species. For this part of the exercise, they consider that an absolute time scale is not available, only stratigraphic ranges. The students zonation schemes are exhaustive, as shown in the zone column of Table 1. The students' identify the nature of each zone (range, concurrent range, or interval) and come to understand the concepts of biozone versus biohorizon and their relationship. The students are then given three assemblages, collected from the same

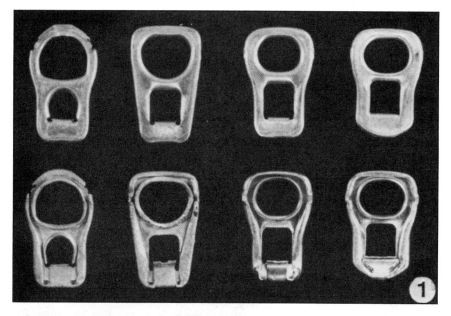

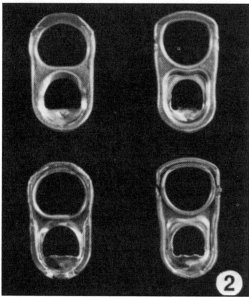

Top: Fig. 1. Dorsal and ventral views of four species assigned to Genus A, Family A, Class A. Bottom: Fig. 2. Dorsal and ventral views of two species assigned to Genus B, Family B, Class A.

region from which the zones were constructed, and are asked to provide an age range based on their biozones. The assemblage in part one would come from Zone VI.

Next, students are given three assemblages, supposedly sent to them from a colleague in Europe who would like to correlate with North American alumenontos biostratigraphy. The students are cautioned that problems may be encountered in correlation over longer distances (for example, absence of data may not be reliable) and are asked to discuss any data problems encountered.

Two nonexhaustive assemblages are provided for identification. For example, an assemblage of Species 4, 9 and 11 would range through Zones VII and VIII. The third assemblage consists of at least two taxa whose ranges do not overlap in this example and an odd specimen not among their 13 morphotypes. The students are expected to recognize from this situation that zonation schemes may not be usable when carried too far from the region for which they were constructed.

Several exercises suitable for an introductory class have been introduced here, but more complex exercises, such as ones dealing with phenetic and cladistic classification, paleobiogeography, or advanced biostratigraphy, could be constructed using alumenontos as a model.

Author Acknowledgment

I would like to thank Cynthia S. Shroba, who recognized the diversity of alumenontos and brought them to my attention; D.B. Blake, who

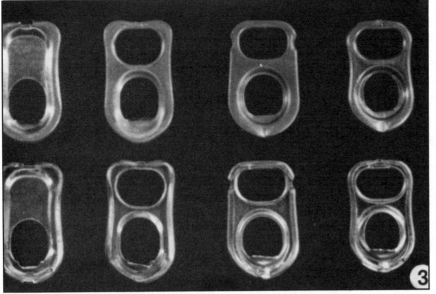

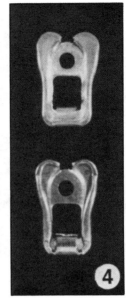

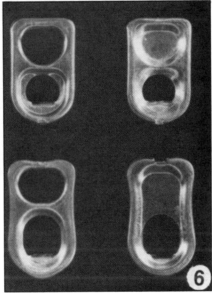

Top left: Fig. 3. Dorsal and ventral views of four species assigned to Genus C, Family B, Class A. Top right: Fig. 4. Dorsal and ventral views of one species assigned to Genus D, Family C, Class B.

Bottom left: Fig. 5. Dorsal and ventral views of two species assigned to Genus E, Family B, Class A. Bottom right: Fig. 6. Dorsal view of four species, displaying homeomorphy between two genera.

read a preliminary draft of this paper; and everyone who has given me alumenontos samples for my collection.

Reference

This activity is based on an original exercise by S.J. Hageman (1989). Use of alumenontos to introduce general paleontologic and biostratigraphic principles. *Journal of Geological Education, 37*(2), 110–13, and is modified and reprinted with permission of the publisher.

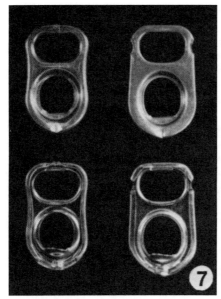

Fig. 7. Dorsal and ventral views of two taxa, which may be interpreted as sexual dimorphs of one species, or as two closely related species.

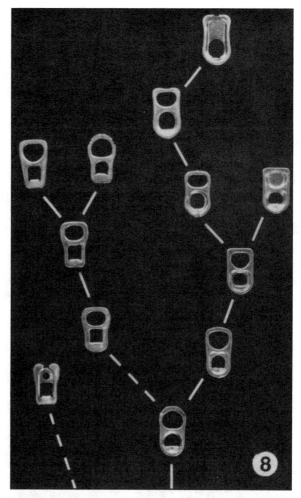

Fig. 8. A hypothetical phylogenetic reconstruction of alumenontos evolution.

A SIMULATION MODEL APPROACH TO THE STUDY OF EVOLUTION

Based on an original activity by
John A. Dawes

In this simulation, students are provided with an assortment of laboratory glassware that they are asked to classify. This proposed classification scheme must be based on some perceived evolutionary trend, such that one piece of glassware is thought to have "descended" from another. Students write their reasons for any classification scheme proposed and draw a chart showing the lines of descent.

Evolutionary Principles Illustrated

- Systematics
- Phylogeny

Introduction

Organisms exist in their present form because they have evolved through time from more primitive ancestors. Therefore, it seems logical to add an evolutionary aspect to the treatment of classification. This would place organisms in their proper context. The following exercise attempts to combine both these aspects of evolutionary biological course work.

Intended Audience

- Life science
- General biology

Materials (for each student group)

- selection of laboratory glassware (20 items), including various sizes of boiling and Erlenmeyer flasks, beakers, etc.. (*Figure 1 below*)

Procedure

It is important that each student group be

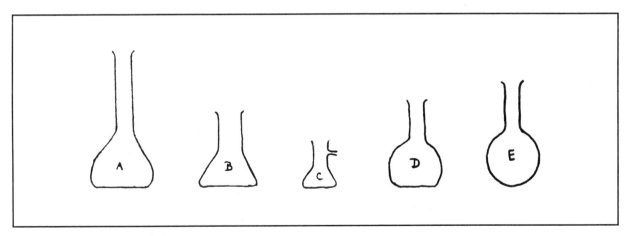

Fig. 1. Sample selection of glassware.

provided with the same selection of laboratory apparatus.

The groups are then set to the task with only the following minimum guidelines to be followed by the student members:

1. Arrange (classify) the selection of laboratory apparatus into logical groups.

a rearrangement of the material in an evolutionary context. Further stages of finer ready movement may follow.

When each group has completed its classification, a written report is compiled giving reasons for the decisions made along the way. If there is not enough time during the session for compilation of reports, then notes, at least, should be

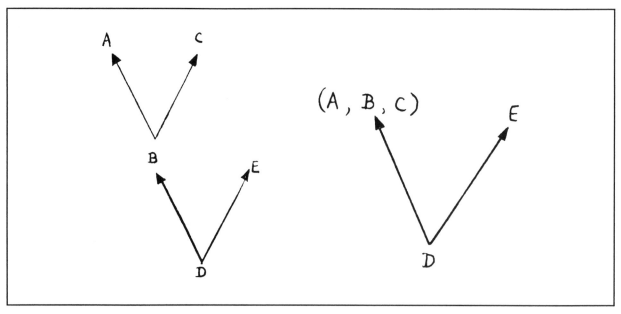

Fig. 2. Evolutionary relationships.

2. Make sure that these groupings reflect evolutionary relationships such as those proposed in Figure 2.

Teams typically first briefly discuss their overall apparatus. The articles of laboratory items, numbering at least 20, are then grouped according to their overall similarities and differences without any consideration of evolutionary relationships.

Next comes a period of readjustment during which subgroups of articles are formed by "splitters," and larger groups are formed through amalgamation of two or more of the original groups by the "lumper." The third stage involves

prepared. The report should be ready in its final form by the next class session.

This next session should be dedicated to discussion, with each group describing its own classification for the class. Students are encouraged to criticize and question the arguments put forward. If each group is given a few minutes to reconstruct its classification of the article prior to the discussion, criticisms and questions can be accurately directed at the sections in question.

The exercise concludes with a summary of the principles of classification in an evolutionary context. This summary is either provided by the teacher or obtained from classroom discussion.

Discussion

Students should be given an opportunity to try other items. A wide selection of textbooks or library books would work. Similarly, writing implements, such as pens (fountain, ballpoint, felt-tipped, quills); pencils (soft, medium, hard); crayons; drawing charcoal; or chalks of various types and colors, would be suitable.

However, laboratory apparatus is probably the most practical material. Storage containers, accessory apparatus (tubing, funnels, beakers or other vessels) are particularly suitable because of the wide range of shapes and sizes available. In making the selection, take care to include some examples of vertical shapes of varying sizes. Whatever choice is given, it is reasonably safe to assume that different groups will arrive at different arrangements.

As a follow-up exercise, ask individuals (or groups) to consult relevant literature and select one organism to study at greater depth. (The natural history section of the local museum or library may prove helpful in this respect.) The end product of this investigation would be a report outlining the evolutionary history of the chosen organisms. Suitable subjects might include the Galapagos finches or giant tortoises, the various species of rhinoceros or zebra, the herring and black-backed gulls *(Larus argentatus* and *L. fuscus),* and the great tit *(Parus maior).* The last two are examples of ring species and provide a wealth of opportunities for investigating evolutionary processes.

Suggested Discussion Questions

1. Which criteria can be used in classification?

2. Is the morphology of an organism more important than other factors?

3. If not always so, when can morphology be considered as of the utmost importance?

4. Cannot similar morphologies be found in unrelated organisms?

5. What is the minimum variation that is considered significant in the separation of organisms into species, genera, families?

6. Can this minimum variation actually be defined?

7. When do different "forms" need to be considered in the classification of organisms?

8. Do ecological/behavioral/geographical factors play any part? If so, how?

9. Can we actually define a species accurately?

10. Can we ever devise a completely natural, rather than artificial, classification?

11. Is it even desirable to propose a natural classification scheme?

Reference

This activity is based on an original exercise by J.A. Dawes (1977). A simulation model approach to the study of evolution. *Journal of College Science Teaching, 7*(2), 102-4, and is modified and reprinted with permission of the publisher.

ILLUSTRATING PHYLOGENY AND CLASSIFICATION

Based on an original activity by
John M. Burns

A model consisting of pieces of hardware (nuts, bolts, screws, etc.) representing animal species in a single phylum may be used to teach problems of taxonomy and the arbitrariness, subjectivity and limitations of higher classifications. Students find this model challenging, stimulating and thought-provoking — it is simple, durable, inexpensive and easy to manipulate. It fits comfortably in one laboratory period, involves a minimum of characters, and presupposes no familiarity with the morphology and accompanying jargon of any particular group of organisms.

Evolutionary Principles Illustrated

• Systematics
• Phylogeny

Introduction

This laboratory exercise is designed to illuminate major problems commonly encountered in the synthetic taxonomic process of determining higher categories. Students are often only dimly aware of problems at this level and, in particular, of the arbitrariness, subjectivity and limitations of the higher levels of classification. Classification schemes themselves are undeniably useful, but are used blindly by many students.

Intended Audience

• Life science
• General biology

Materials (for each student group)

• set of approximately 30 assorted pieces of hardware, such as nuts, bolts, etc. (See Figure 1 on page 121.)

Procedures

Part I

Each group of students will receive an envelope (marked "Classification") containing the assorted pieces of small hardware. (Do not mix the contents of any two envelopes.)

Each object represents a different species of animal. Have students study these "organisms" carefully, comparing each species with every other species to detect similarities and differences among them. Use all available "taxonomic characters" to work out the relationships between the organisms and to arrange them in an orderly hierarchic scheme that more or less reflects these relationships. Assume that all of these animals belong to a single phylum and limit your classification of them to the following taxonomic categories: class, order, family, genus, species. It is a good idea to avoid finer subdivisions such as superorder of subfamily.

Some species may be considered more primitive than others and perhaps directly ancestral to others. (Hence, some of the "specimens" with which you are working may be "fossils" rather than recently collected specimens of living

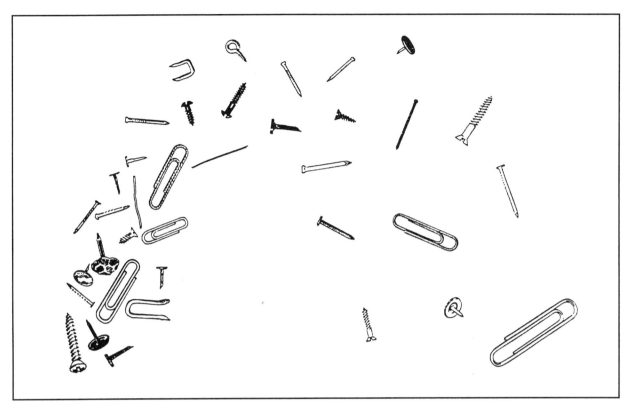

Fig. 1. A typical first look at the various "species" of this model.

species.) You will have to make considerable use of linear measurements as characters. Assume that there is no appreciable variation within each species.

After contemplating the entire problem, students prepare a formal presentation of the interpretations, as follows:

1. Make a phylogenetic diagram of all the species.

2. Make a classification embracing all the species. As you establish this classification, give the characteristics of each taxonomic group.

3. Steps 1 and 2 are intricately related. Justify both your phylogenetic arrangement and your classification by explaining, for example, why you have judged some forms to be more primitive and others more advanced, why you have

seen fit to make the groupings that you have, and so forth.

4. Discuss the following questions briefly:

a) How is phylogenetic relationship inferred from morphology?

b) What difficulties arise in the process of translating a phylogenetic diagram into a classification?

c) Is information lost in this process?

d) What arbitrariness, if any, is inherent in the practice of modern evolutionary taxonomy and classification?

5. Construct a dichotomous key (which can be artificial rather than natural) allowing ready and rapid identification of each species.

Students are often astounded when they first pour out their hardware (Figure 1). Some laugh. A few may stare blankly at the parts for minutes on end. Such individuals may need comments or questions to get started.

In general, the best procedure is (1) to leave the students alone, letting each pour over his/her hardware and ponder his/her own phylogenetic arrangement and the tentative groupings he/she would make in shifting from phylogeny to classification; and then (2) to encourage the students to compare notes. They are usually surprised at how much their interpretations differ and frequently get into heated but healthy arguments.

Students may disagree strongly — much like professional taxonomists — about what is primitive and advanced (and what these terms mean), about what direction an apparent sequence takes (e.g., small to large or the reverse), about what could conceivably give rise to what — in short, about most of the relationships they are trying to determine.

Occasionally, students may ask — in real anguish — if some characters are more important than others? For example, are differences in color or size as trivial as they often appear to be? Is it true with living organisms, as it seems to be here, that what are good characters in one group are not necessarily very helpful in another?

Some students are disturbed to learn that there is no one correct solution, that the exercise is anything but black and white, and that (given the information in the various "species" and the rules of the model) many interpretations are acceptable. Most realize that, despite this, there are numerous arrangements and groupings that are plainly indefensible. Almost all come to discern a series of parallels between the model and the biological situation.

The model not only emphasizes the difficulties

and differences of opinion among workers in establishing a phylogeny with incomplete data at hand (the standard condition) but also drives home the ways in which information is lost in going from an evolutionary diagram to a hierarchic classification. Students grapple directly with such conflicting forces as splitting versus lumping and, more importantly, vertical versus horizontal classification and are often severely distressed when they recognize that, in some instances, they cannot avoid rather arbitrarily placing related species in different major groups.

Even if two students agree on a phylogeny, they may yet produce different but valid classifications consistent with that phylogeny. For example, some students want to discard the screw eye. Some of those who perceive that the screw eye will not readily fit in their phylogenetic scheme are brought face-to-face with the concept of convergence. They find it easiest (but still not altogether satisfying) to suggest independent origin of threads in the screw eye and the screws.

Most students find the exercise challenging and imaginative. They usually emerge from it more appreciative of the difficulties of practicing taxonomy, somewhat disillusioned with and bothered by classification, aware of many of the limitations of our system, and prepared to take classifications, in the future, with a grain (at least) of salt.

Part II

In Part I, the students were asked to assume that appreciable intraspecific variation is nonexistent. This — like many other features of this exercise — is a huge oversimplification.

Take one of the envelopes marked "Samples." It contains samples of 21 "adult males" of each of two "species" collected at the same time and place. (Individuals of both species are represented here by wires; the color of the wire serves to distinguish the two species.) Assume that each sample perfectly reflects the population from which it was drawn, and further assume that each population was exposed throughout its development to identical environmental conditions so that individual variation in the sample stems from genetic variation in the population.

Measure each individual in both samples to the nearest half-centimeter, record the measurements, calculate the mean length for each sample, compare the two means thus obtained, and then plot the frequency distribution for each sample in the form of a bar graph, with length increasing to the right along the abscissa and number of individuals increasing upward along the ordinate.

Suggested Discussion Questions

Place the resulting histograms one above the other (see Figure 2).

1. On what basis, other than color, can you distinguish the two species? If you had only one or two specimens of each species, could you tell them apart using this character? Explain.

2. If the environment in which these two species live should undergo relatively rapid change,

which species might have the better chance of surviving? Why?

Reference

This activity is based on an original exercise by J.M. Burns (1972). A simple model illustrating problems of phylogeny and classification. *Systematic Zoology, 17*(1), 170–173, and is modified and reprinted with the permission of the publisher.

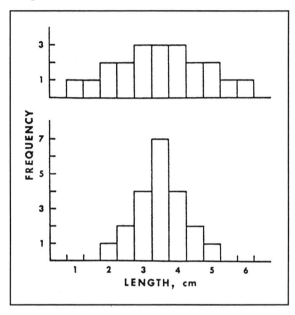

Fig. 2. Variation in "body" length in samples of each of two "species" of wire.

THE CAMINALCULE FAMILY TREE

Based on an original activity by
D.J. Smith

This simulation involves studying drawings of members of an imaginary phylum of animals called "Caminacules." Students are presented a series of open-ended problems in evolutionary biology and taxonomy, starting with an investigation of some of the basic principles of classification. The lack of definite relationships between the different evolutionary lines means that any hypotheses in agreement with the data may be considered valid.

Evolutionary Principles Illustrated

- Phylogeny
- Systematics

Introduction

The Caminalcules are members of a phylum of imaginary organisms invented by the late Dr. J.R. Camin of the University of Kansas as the basis for a series of advanced exercises in numerical taxonomy, as described by Sokal (1966).

"Caminalcules have been found to be a useful device for teaching aspects of taxonomy and evolutionary biology because they are completely hypothetical and can be custom made to suit the requirements of a particular exercise. In addition, a restricted range of characteristics can be displayed, so that the amount of information available to students may be controlled, and they have

been found to be amusing to students across a large range of age and ability."

Intended Audience

- General biology
- Advanced biology

Materials (for each student group)

- set of Caminalcule drawings
- scissors (optional)

Procedure

In developing a series of "caminculoids" for school use, no attempt was made to follow Sokal's rather advanced treatment or to apply the strict rules of design originally suggested by Camin. Six basic forms were drawn, as shown in Figure 1. They represent some body plans appropriate to life on land, in water, and in the air.

From these basic six types, new forms were designed along what seem to be biologically consistent lines to represent various evolutionary sequences. Members of the complete set of caminculoid figures may be presented to students, each glued to a card or on a single sheet (Figure 2).

Part I: Proposing Biological Names

We give things names in order to be able to

describe them more easily. For example, the word "pig" saves us having to give a very long description every time we want to talk about that animal.

Provide students with an example of each of the six Caminalcule forms shown in Figure 1. Pretend that each one occurs near where the students live. Have students invent suitable names for each organism. See how many people in the class can recognize which name goes with which organism.

People try to give names to everything they see around them, but different people living far apart often give totally different names to exactly the same thing. For example, the names "king cat," "ghost cat," "catamount," "panther," "puma," "cougar," and "mountain lion" all refer to the same animal.

This particular exercise proceeds from the uncertainty of common names to the need for a systematic nomenclature. The unwieldiness of the descriptive names used by the early classifiers is contrasted with the simplicity of binomial names. For example, the "carnation" was originally described as *"Dianthus floribus solitariis," "squamis calycinis subovatis brevissimis," "corollis crenatis,"* or by the Linnaean binomial *Dianthus caryophyllus.*

Ask students to invent suitable binomial names for the six sample Caminalcules presented. Rival binomial names provide a good arena for discussion about precedence and other taxonomic conventions. For more advanced students, reference may be made to discussions regarding precedence in taxonomic papers in journals: for example, the taxonomic review in Higgins (1974).

This stage of the exercise ends with the general acceptance of definitive binomial names for the original six organisms. The names are retained throughout the rest of the study.

Part II: Proposing Phylogenies

The question of relationships and classification arises when students are asked to derive names for the whole set of Caminalcules (see Figure 2 on page 127) while retaining those agreed upon for the original six. It becomes necessary to consider how similar-looking organisms may be related and how these relationships may be referred to in sets and subsets. The scientific cards' inadequacy of descriptive information soon becomes apparent to most students. This means that the students must try to decide for themselves what and how much information is necessary for realistic classification.

The question of how many distinct types of organisms are represented among the 29 different pictures in the set necessitates some discussion of natural variation and polymorphism.

Fig. 1. The six main Caminalcule forms: A. squid; B. snail; C. flirt; D. parasite; E. generalized land; and F. generalized aquatic.

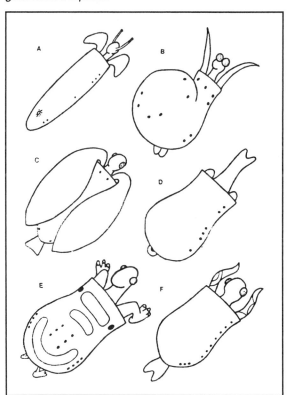

Some students realize that there is no way of telling whether the Caminalcules are sexually dimorphic.

The value of field studies in elucidating problems of this kind can be stressed. Bird study has been found to be helpful here because, in the case of highly similar animals, such as sparrows, only close and careful observation will reveal the true pattern of relationships. This is an example immediately available for study by students. The aim of this part of the exercise is to investigate some of the basic ideas of taxonomy and to establish the need for much detailed information.

At a fairly advanced level, consideration of polymorphism and variation leads fairly directly on to the subject of evolution, and the Caminalcules offer good scope for treatment of this topic. At the upper end of school and beyond, students who are asked to sort the cards into putative evolutionary lines will generally notice a degree of coincidence between these and their taxonomic sets and subsets; though with a monolithic taxonomy, it is difficult to ensure that this is not merely a restatement of the classification.

In the absence of a clear ancestral type, several parallel arguments may be advanced. When rival groups are asked to defend their positions, a hardening of opinion tends to be seen, with students taking great exception to opposition, even in the face of a generally agreed upon inadequacy of information. This has been found to be a suitable point to introduce discussion of the evolutionary debates of the last century, and a consideration of the relative status of different evolutionary ideas that are not amendable to empirical investigation.

References

The article from which this activity was taken was written by D.J. Smith, but permission to reprint the "Caminalcule" drawings was granted by the noted biologist R.R. Sokal, who was a colleague of the late J.R. Camin, developer of the Caminalcules.

This activity is based on an exercise by D.J. Smith (1975) and was originally titled "Simulation in taxonomy: The use of Caminalcules." *Journal of Biological Education, 9*(3/4), 155–157, and is modified and reprinted with the permission of the publisher.

Fig. 2. Various types of Caminalcules.

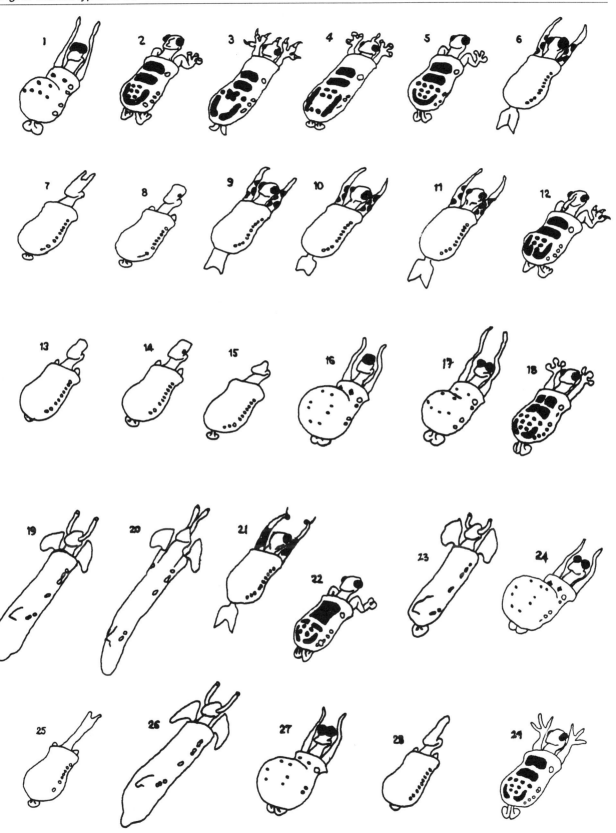

IX. THE NEW EVOLUTIONARY SYNTHESIS

Darwin and Wallace certainly did not solve all of the problems associated with biological evolution with their proposed mechanism of natural selection. After all, neither could explain the source of the variation central to the concept of natural selection. Work during the century since their discovery has added much to our store of knowledge about descent with modification, but rather than threatening the work of these two pioneers, the central tenets laid down so long ago have held firm.

New discoveries in genetics, the nature of mutation, and innovative ideas about the rate of evolutionary change have initiated the period known as the *new synthesis* of evolutionary biology. This chapter contains several activities focused on aspects of descent with modification unknown to Darwin and Wallace.

In the first activity, students examine simulated data discovered from two mythical creatures: one illustrating a gradualistic evolutionary sequence and the other the new punctuated progression. In the punctuated example, the form of the creatures remains essentially unchanged (equilibrium) for a long time period and then suddenly changes (punctuation). For many organisms, this punctuated equilibrium style of evolution may be much closer to an actual representation than the gradualism advocated by scientists previously.

The final exercise in this monograph uses the computer to simulate macroevolutionary change. Some have criticized evolution as impossible because of the randomness inherent in most models of evolutionary change. Opponents of evolution have said that, if text were randomly generated, not even one sentence of Shakespeare would ever be generated — the odds are simply too immense. This argument against evolution falls apart with the simulation provided as the last activity in this section. If the useful mutations are preserved, random selection for those unhelpful characteristics — in this case, letters — will produce a work of Shakespeare more quickly than most evolution-dissenters would like to believe.

MODELING MODES OF EVOLUTION: COMPARING PHYLETIC GRADUALISM AND PUNCTUATED EQUILIBRIUM

An original activity by
William F. McComas and Brian J. Alters

This activity provides students an opportunity to explore the tempo and mode of evolution by analyzing data and constructing two evolutionary trees, one gradualistic and one punctuated. The data are fictitious, as are the creatures used as illustrations, but are representative of real data.

Evolutionary Principles Illustrated

- Tempo and mode of evolution
- Determination of speciation

Introduction

"Paleontologists have discovered two major patterns in life that make it difficult to support a totally uniformitarian view of life's development" (Benton 1993, p.100). These two views are known as *phyletic gradualism* and *punctuated equilibrium.*

Phyletic gradualism is the traditional Darwinian view that an interminable number of intermediate forms have existed, linking together all species in each group by gradations as fine as our existing varieties (Darwin 1975).

Punctuated equilibrium, developed by Niles Eldredge and Stephen Jay Gould (1972), offers a contrasting view that organic evolution is not steady and regular but episodic and jerky, with long periods of small changes interspersed with rapid bursts of large-scale transformation of species. The latter pattern explains that the

"gaps" in the fossil record are not simply missing data that will show up some day — as maintained by gradualists — but are real and must be interpreted as such.

Intended Audience

- General biology
- Advanced biology

Materials (for each student group)

- copies of the Caminalcules in the *Genus Molluscaformis* and in the *Genus Pedivarious* (different colors of paper will be useful)
- copies of geologic columns for the sites where samples were found (*If enlarged 135%, these charts will fit neatly on legal size paper.*)
- scissors
- graph paper (optional)

Procedure

Each student group should have photocopies of both the Caminalcule genera and the accompanying strata sheets. The students should cut out all the Caminalcules, keeping the related data attached. Each Caminalcule provided represents the morphological average of a number of Caminalcules found at a particular location.

The "average of" number is located below each Caminalcule in parentheses. For example, one Caminalcule might be represented by an average of four finds. This information, although fictitious, is provided to help the student understand that there is some morphological variation within the specimens found at a given site and that conclusions are based on a range of specimens rather than on a single individual.

The name listed with each Caminalcule is the name of the formation or layer in which it was found. If the specimen is listed as "Upper Wallacian," it was found in the upper, or more recent part, of the layer called the "Wallacian Formation." At the left of each stratigraphic column are numbers representing the number of thousands of years that it took to form that particular layer. Following a discussion of the issues mentioned here, students should follow the specific instructions below.

Specific Instructions

1. Working with one genus at a time, each student group should arrange the Caminalcules on the appropriate stratigraphic column by placing each individual in the stratum (layer) in which it was found. *(The figures noted below appear on pages 136-140 following this activity.)*

2. Next, the species in the genus should be arranged into a logical morphology versus time tree (Figure 1). *Note: It is best if the students do not see these example trees prior to constructing their own.*

3. Draw the genus evolution tree on a morphology versus time axis (Figure 2). Place the correct time units on the Y-axis. Morphological change will have to be estimated (no units).

4. Repeat the previous steps for *Genus Pedivarious* (see Figures 3-5). The *Genus Pedivarious* tree should look like Figure 3.

5. To understand punctuated equilibrium, one must examine it point-by-point with Darwin's view of phyletic gradualism. Have the students make a comparison list of the two trees. The two patterns of evolution along with implications for each are contrasted in Table 1 (see page 141).

6. Have students define the following with reference to their proposed trees:
• Transformation
• Speciation
• The geological meaning of "fast" and "slow"
• Evidence
• The question of how paleontologists decide if organisms are of different species
• Lineage of descent with modification
• Strata
• Morphology

Discussion

After making the basic comparisons of phyletic gradualism and punctuated equilibrium, divide the class into two groups. Students should read some of the background materials detailing the scientific merit of each evolutionary pattern.

General review
• B.J. Alters and W.F. McComas (1994).
• R. Lewin (1980).

Pro
• S.J. Gould (1977, 1991).

Con
• P. Whitfield (1993).
• E.O. Wilson (1992).

Important Considerations:

• Fossils may be broken, distorted and/or have parts missing.

• Only 10% of geologic time is available in sedimentary layers (Van Andel 1981).

• Paleontologists generally decide if fossils are of differing species by comparing them to similar living organisms.

Debate and/or Discussion Topics:

Some phyletic gradualists would state that the nine layers in the *Pedivarious* evolutionary sequence are not complete (Figure 4). Maybe little or no rock was formed in a period between Gouldian and Eldredgean, and consequently there are no fossils represented from this period. Therefore, the actual evolution of the *Genus Pedivarious* could be gradual!

Punctuationalists would counter by stating that the gradualists are arguing from lack of evidence. (This would be a great place to have a discussion about the nature of science, such as: What counts as scientific evidence?) As Gould and Eldredge (1977) state, "Phyletic gradualism was an *a priori* assertion from the start — it was never 'seen' in the rocks ... we think that it has now become an empirical fallacy" (p. 115).

Author Acknowledgment

We acknowledge the contribution of the late J.R. Camin of the University of Kansas who developed the fictitious organisms called "Caminalcules." He applied basic evolutionary principles and designed these creatures to be used in teaching various aspects of evolutionary biology. We have modified two of the Caminalcules for use in the activity presented here.

We would also like to thank Susan Lafferty, science education specialist of the Los Angeles County Museum of Natural History, for lending her artistic talents by drawing the modified Caminalcules.

Reference

This is an original activity by W.F. McComas and B.J. Alters (1994). Modeling modes of evolution: A comparision of phyletic gradualism and punctuated equilibrium. *The American Biology Teacher* 56(6), 354-360, and is modified and reprinted with the permission of the publisher.

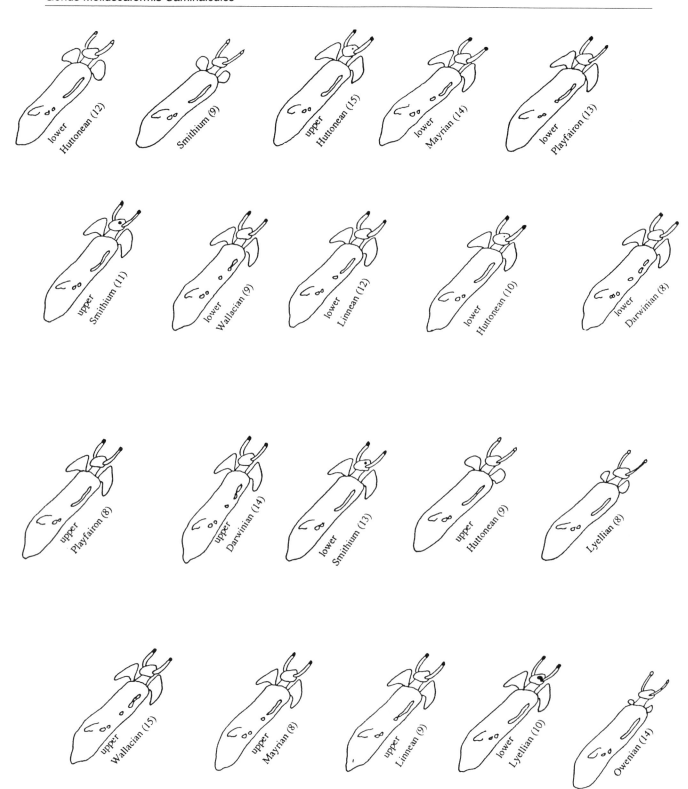

upper
Eldredgean (10)

upper
Mullerian (15)

upper
Huxlian (11)

lower
Huxlian (9)

Aristotelian (13)

lower
Eldredgean (13)

upper
Gouldian (8)

lower
Mullerian (15)

Mendelian (9)

lower
Mendelian (15)

Simpsonean (14)

Raupian (11)

lower
Eldredgean (13)

lower
Gouldian (14)

upper
Eldredgean (10)

Lamarckian (8)

Fig. 1. A completed chart showing the placement of members of the Genus Molluscaformis arranged by morphological characteristics and the layer in which each sample was found.

STRATIGRAPHIC SEQUENCE FOR THE GENUS MOLLUSCAFORMIS

Formation	Duration	
Owenian	250,000	
Lyellian	300,000	
Smithium	150,000	
Huttonean	100,000	
Playfairon	80,000	
Linnean	175,000	
Mayrian	200,000	
Wallacian	50,000	
Darwinian	100,000	

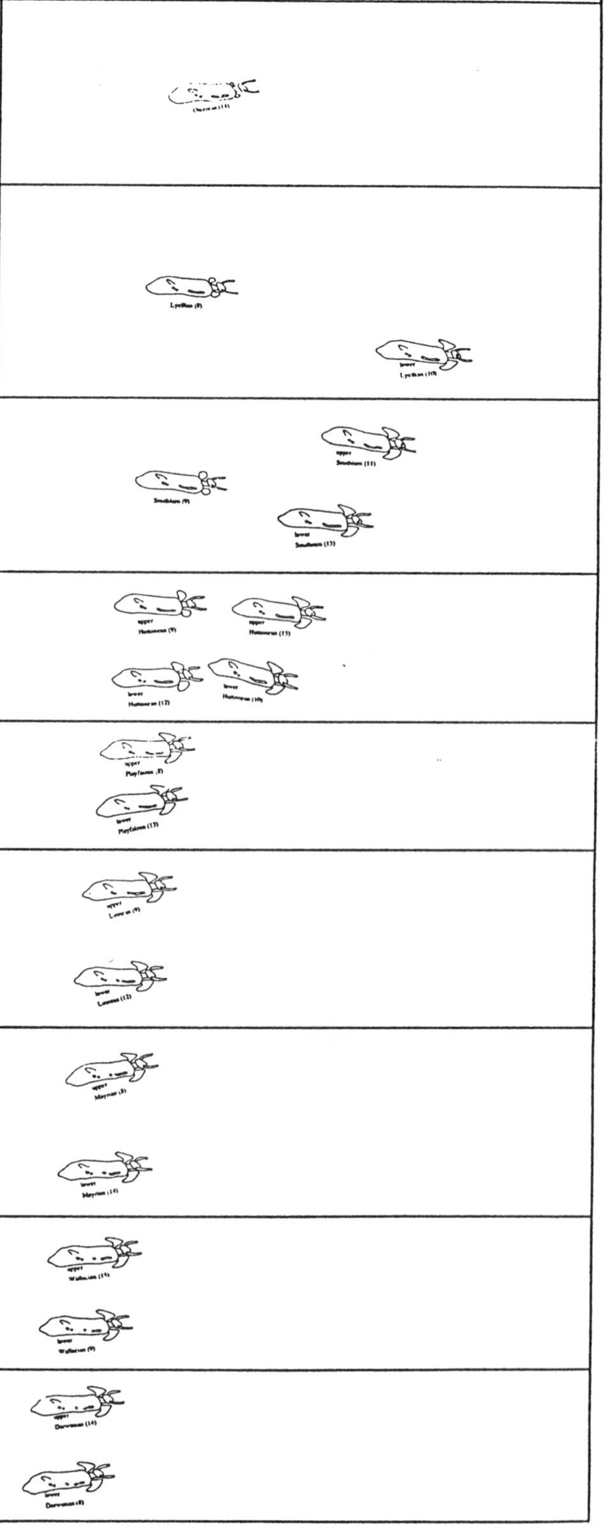

GENUS *MOLLUSCAFORMIS*

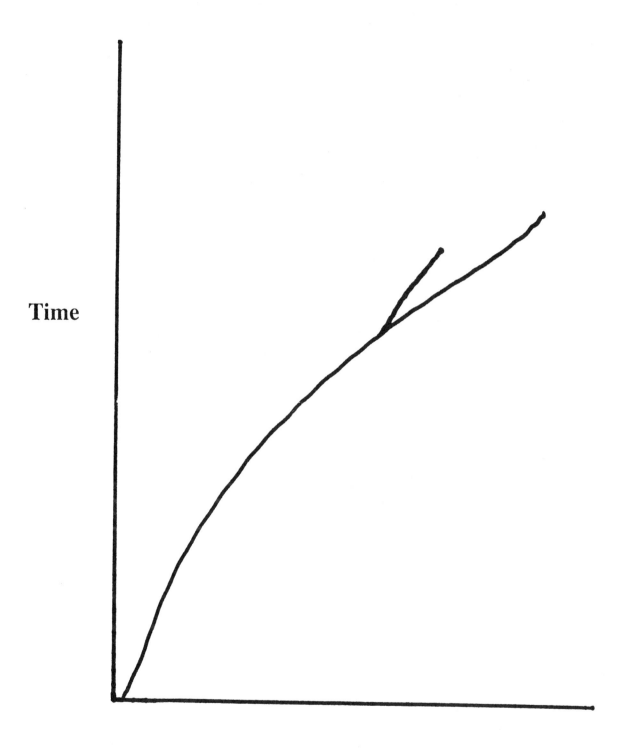

Morphology

Fig. 3. A completed chart showing the placement of members of the Genus Pedivarious arranged by morphological characteristics and the layer in which each sample was found.

STRATIGRAPHIC SEQUENCE FOR THE GENUS *PEDIVARIOUS*

Formation	Duration	
Lamarckian	75,000	
Aristotelian	100,000	
Simpsonean	200,000	
Mendelian	100,000	
Eldredgean	35,000	
Gouldian	50,000	
Huxlian	250,000	
Raupian	100,000	
Mullerian	200,000	

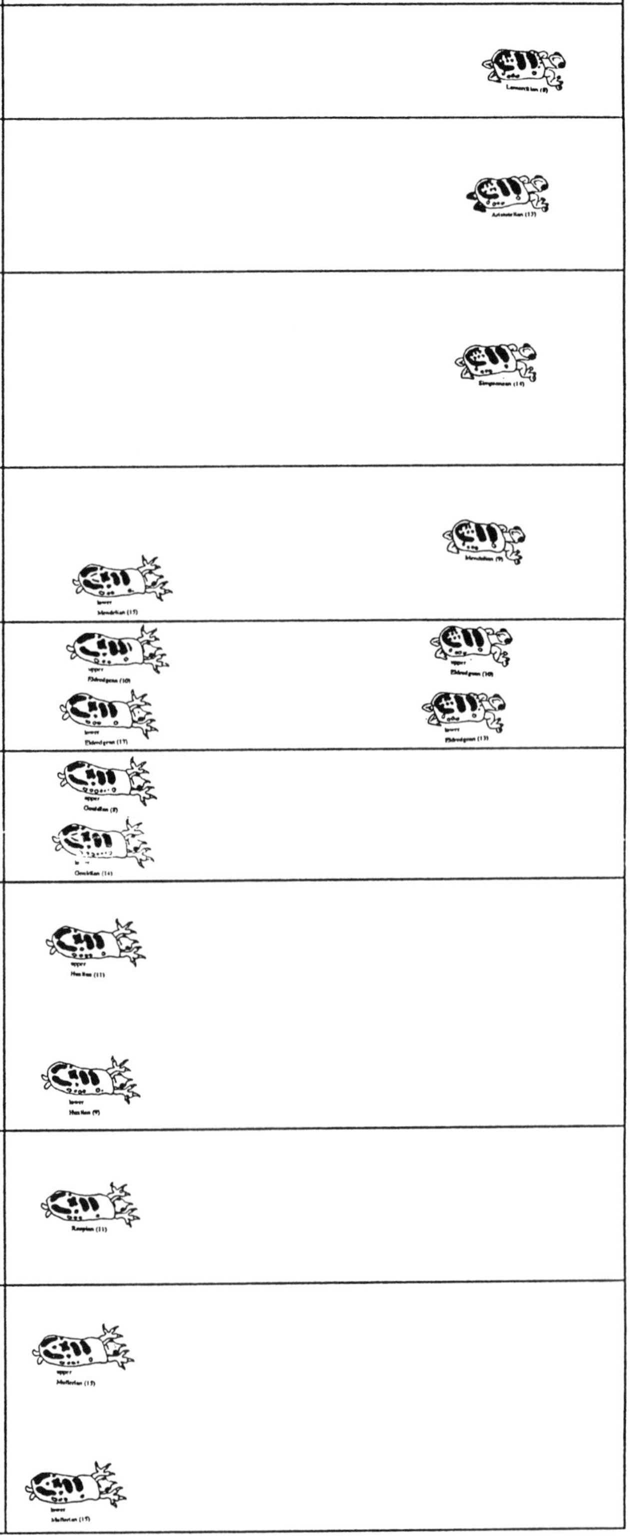

GENUS *PEDIVARIOUS*

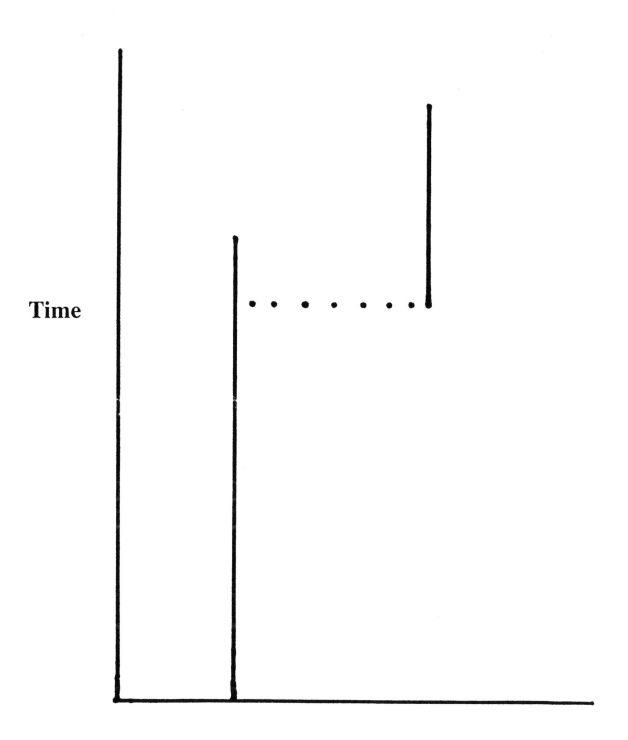

Fig. 5. Geologic column illustrating a possible erosional event that provides support for phyletic gradualism in the case of the Genus Pedivarious.

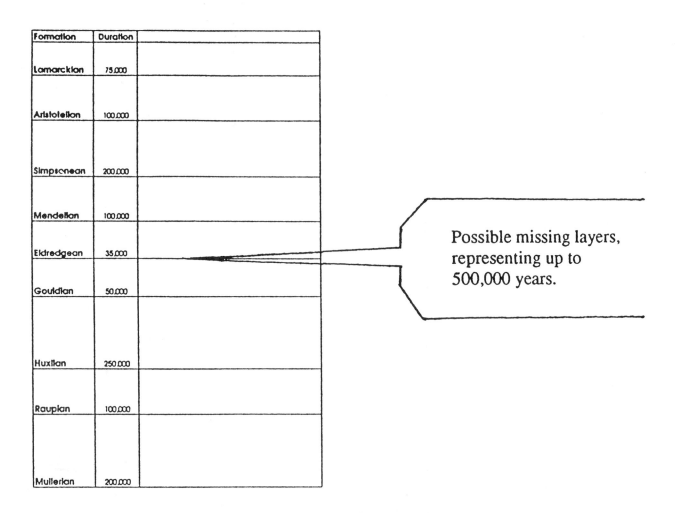

Formation	Duration	
Lamarckian	75,000	
Aristotelian	100,000	
Simpsonean	200,000	
Mendelian	100,000	
Eldredgean	35,000	
Gouldian	50,000	
Huxllan	250,000	
Rauplan	100,000	
Mullerlan	200,000	

Possible missing layers, representing up to 500,000 years.

Table 1. Comparison of phyletic gradualism and punctuated equilibrium.

Caminalcules	
Fictional Genus: *Molluscaformis*	Fictional Genus: *Pedivarious*
Name: Phyletic Gradualism	Name: Punctuated Equilibrium
Principal Proponent: Darwin	Principal Proponents: Eldredge and Gould
• New species develop gradually and slowly with little evidence of stasis (no significant change)	• New species develop rapidly and then experience long periods of stasis
• The fossil record should contain numerous transitional forms within the lineage of any one type of organism	• The fossil record should contain few transitional forms with the maintenance of given forms for long periods of time
• New species arise via the transformation of an ancestral population	• New species arise as lineages are split
• The entire ancestral form usually transforms into the new species	• A small subpopulation of the ancestral form gives rise to the new species
• Speciation usually involves the entire geographic range of the species (called sympatry)	• The subpopulation is in an isolated area at the periphery of the range (called allopatry)

Adapted from Eldredge, 1989; Futuyma, 1986; Rhodes, 1983; and Gould & Eldredge, 1977

A COMPUTER SIMULATION OF THE EVOLUTIONARY RATE IN MACROEVOLUTION

Based on an original activity by
O.B. Marco and V.S. López

This simulation provides a view of the controversy with respect to the mode and tempo of evolution by proposing a simple model that can help students to think about evolution and better understand the gradualist and punctuationist macroevolutive approach.

The model affords students an easy and friendly teaching tool to introduce these concepts and other related ones (i.e., mutation, selection, fitness, extinction, origin of life, etc.) in the classroom. One of the greatest difficulties in teaching these issues is the inability to perform experiments concerning evolutionary predictions. Computer programs presented here can overcome some of these difficulties.

Evolutionary Principles Illustrated

• Mode and tempo of evolution

Introduction

Presently, there are some controversies among scientists who have accepted the theory postulated by Darwin (enhanced in the present century by neo-Darwinian contributions) but disagree with some of its aspects (Ridley 1985). Among these controversies are the level of selection (species, individuals or genes), the power of natural selection, the neutral theory, macroevolution, etc.

A major controversy is over the rate of evolution and when most significant evolutionary change occurs. Here there are two opposing schools: phyletic gradualism (Dawkins 1987) and punctuated equilibrium (Gould & Eldredge 1977).

Intended Audience

• Advanced biology

Materials (for each student group)

• simulation software (Appendix A, p. 150)
• compatible computer

Procedure – The Model

The model presented here is a very simple producer of random sentences. It is analogous to the one used by Richard Dawkins in *The Blind Watchmaker* (1987) to explain the difference between single-step selection and cumulative selection. Dawkins starts from Hamlet's sentence, "Methinks it is like a weasel," and designs two different computer programs to obtain it from a random set of characters with the correct length (see Figure 1).

1. Single-step selection of random variation begins by typing a random sequence of 28 characters (the length of Shakespeare's sentence) and comparing it to the target phrase, "Methinks it is like a weasel." If the phrase is typed correctly, the experiment ends. If the sentence typed is different from the target, it makes another trial of 28 characters, and so on.

Model:
"METHINKS-IT-IS-LIKE-A-WEASEL"

Single-step selection of random variation

Cumulative selection of random variation

Random sequence of
28 characters

Random sequence of
28 characters
(mother-phrase)

duplication process with a
random copying error

daughter-phrase

does it match the
model exactly ?

does it match the model
better than mother-phrase?

NO   YES

YES   NO

daughter-phrase assumes
the role of mother-phrase

does it match the
model exactly ?

END
(target reached)

YES   NO

2. Cumulative selection of random variation begins again by typing a random sequence of 28 characters, but now the first nonsense sentence is duplicated repeatedly with a certain chance of random error in the copying. The computer examines the "mutant" phrase as well as its ancestors and chooses the one which, however slightly, most resembles the target, "Methinks it is like a weasel." The closer phrase plays the role of pattern in the next copying, and this goes on generation after generation.

It is easy to calculate how long we should reasonably expect to wait for the single-step selection process to type "Methinks it is like a weasel." The probability of a trial of 28 random correct blows at a keyboard with 27 keys (26 letters and a space bar) is extraordinarily small $1/27$ to the power 28, approximately 8.35×10^{41}.

With this chance, we expect, with a computer that generates 100 random phrases per second, to reach the target in a time that, compared with the age of the Universe, makes the latter negligible. On the other hand, the second process, the cumulative selection, takes a few seconds or minutes, depending on the computer language used, to reach the objective.

Dawkins uses this example to explain the difference between single-step and cumulative selection of random variation, proving that creationist arguments about evolution being a "random process," and thus impossible, fall into severe error. If selection is invoked, then even if variation is randomly generated, it can very quickly be formed into adaptive patterns. Moreover, he warns that the model:

"… is misleading in important ways; one of these is that in each generation of selective 'breeding,' the mutant 'progeny' phrases were judged according to the criterion of resemblance to a distant ideal target, the phrase 'Methinks it is like a weasel.' Life isn't like that. Evolution has no long-term goal. There is no long-distance target, no final perfection to serve as a criterion for selection, although human vanity cherishes the absurd notion that our species is the final goal of evolution … the watchmaker that is cumulative natural selection is blind to the future and has no long-term goal." (Dawkins, p. 50)

In real life, species evolve through locally established criteria.

Taking into account these reflections on Dawkins' model, we have changed the role of the target phrase in our model: "Methinks it is like a weasel" is not a final goal to reach but represents a set of attributes of the different generations of computer sentences to survive.

Our model includes the existence of organisms whose most important variables (a_i, b_i, c_i,...) must be compatible with the characteristics of the environment where they live (A, B, C, ...). To survive, the living beings have a set of characteristics responsible for their adaptation to the environment: characteristics with a finite and discreet range of possibilities (the small letters of the alphabet plus the spaces between words). We consider all possibilities that allow the existence of viable organisms as the maximum variability, and the particular existence of each possibility is due to random "mutation" and "natural selection." The feasible rhythm of change is constant in the model. A mutation is produced by each phrase's generation, but this kind of change is not necessarily an improvement in adaptation.

Furthermore, in our model there exists a certain combination of characters (a_o, b_o, c_o, ...) representing the best adaptation to the environment. Therefore, the appearance at random of an "a" value, corresponding to the characteristics "a," is evaluated by "natural selection" in relation to the rest of "a_i" values. The distance between the present configuration and the best adaptation to the environment (ao, bo, co,...) is given by the distance:

$$\mathbf{DIS} = |\mathbf{ao-ai}| + |\mathbf{bo-bi}| + |\mathbf{co-ci}| +.$$

A numerical value (from 1 to 27) is assigned to each characteristic so that the distance (DIS) contains information about the adaptation level of a sentence from the adaptation level of each of its own characteristics.

At this moment, we want to establish clearly two questions aroused by the proposed model. The first one is that, in our model, random drift is not connected with punctualism. Genetic drift is important in speciation models compatible with punctuated equilibrium — i.e., Wright's shifting balance model (Dobzhansky et al. 1986).

The reason that excludes drift from the model leads us to the second question. We have used the term "organism" throughout this paper when "population" is meant. Organisms do not evolve. However, we have maintained the term "organism" because the structure of our populations is quite special. All organisms that constitute each population are identical. Obviously, genetic drift is excluded from the model.

Experiments Using the Model: The Role of the Environment

We have worked with a set of 28 characters whose best adaptation is the sentence "Methinks it is like a weasel." We started from a random configuration with the correct character length: "cdozhmhyeloucuoxmqftvgxekcsx." We bore in mind that it is unlikely that a random sequence of letters could "survive" in the environment (random DNA does not make functional products). But we can consider this starting point as similar to that probably produced when the first living form emerged from inanimate material.

It is feasible to imagine a situation with soft selection where all sequences can survive, but the ones closer to the environmental characteristics survive better. Having overcome this

trouble, this "ancestor" evolved according to the proposed model rules. Some of its descendant combinations were:

cdtimhyelocuoxmqftggxekisl
   generation 25
cdtihmhselohcuonmqfobgxegiml
   generation 74
ndtihmhselohjulnmqfobgxegiml
   generation 103
ndtihmhsblocjifdbdxebrfl
   generation 202
metihmhsbiucis–kjif–b–wearfl
   generation 445
metimgsait–is–kjke–a–wearfl
   generation 700
methimgs–it–is–kike–a–weasel
   generation 1526

If we graph the number of adaptation improvements against time units and their speed of appearance, we obtain the curve represented in Figure 2.

Note that the rate of the adaptation is not constant. These results are very similar to those obtained by Haldane and Bader in their studies

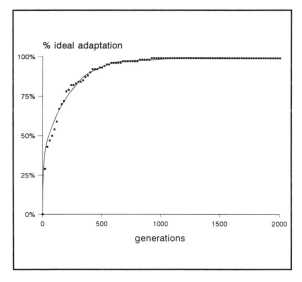

Fig. 2. The level of adaptation in a lineage through time in a constant environment. The experimental points and their best fit are shown.

on the relative rhythm of change in the characteristics of fossils (Simpson 1983). This kind of result is logical, because when the adaptative distance (DIS) is decreasing, the probability that the next random adaptation is an improvement decreases quickly. The shape of the adjusted curve shows two distinctive sections: the first one with a great slope and the second one asymptotic.

The question, "Do lineages evolve at different rates in different times?" is very important. The punctuated equilibrium theory suggests that evolution has a nonconstant tempo, with short intervals of fast evolution, accompanied by speciation processes, interrupted by very long periods with no evolutionary change. The phyletic gradualism asserts that the evolutionary rate is nearly constant in time.

The facts studied up to now in the fossil record have not settled which of the two theories is the more correct; some studies are in agreement with gradualist theory but others with punctuationist statements (see Chapter 9 in Ridley 1985). In our model, the "fossil record" has no gaps, is fully complete, and evolutionary rates of the letter sets are closer to punctuated equilibrium theory than gradualist theory, as can be seen in Figure 2.

This does not deny the existence of periods with gradual evolution, with a nearly constant rate of change. In fact, both theories result from the same mathematical model, and it is not necessary to consider them as in opposition. Both phenomena have the same cause, environmental change, but depending on the kind of change, the response of the evolution rate is different. If the environment is slowly and gently modified, organisms will respond with a gradual evolutionary rate (changes made by random mutation and natural selection); but if environmental changes are sharp and large, the speed of change will increase spectacularly and sudden changes will appear (Stewart 1990).

Let us return now to our model to test these statements. In order to ascertain the influence that environment exerts on the evolution rate, we have modeled the simile used by Stephen Jay Gould in his book *Wonderful Life* (1989). We have rewound the tape of life to record it again from the same starting point, as Frank Capra did in his famous film, "It's a Wonderful Life."

We then repeated the previous evolutionary experiment, which took place in a constant environment, by allowing environmental changes so that more than 2,500 generations of the initial environment "METHINKS–IT–IS–LIKE–A–WEASEL" are converted into the first sentence from Don Quixote, "–EN–UN— LUGAR—DE–LA– MANCHA."

The transformation of Shakespeare's sentence into Cervantes' sentence was modeled in two different ways. The first goes through a series of small gradual changes (always less than 10% of the maximum feasible change, measured by DIS). For the second, these gradual changes have two major sudden changes inserted (near 50% of the maximum feasible change, measured by DIS), change that we will call "catastrophic." The percentages of character changing in time are given in Figure 3.

Once these two dynamic environments were designed, we rewound the tape from the first experiment, made in a constant environment, and recorded it again with the evolution that takes place in every new environment. The starting point chosen by us to begin the new evolutionary sequences was generation 700, because this generation gave us a sentence with enough adaptation level to be considered viable in the environment where it is evolving. At this moment of the evolutionary process, we consider that competitive phenomena are sufficient for producing a stronger selection pressure.

We have measured the percentage of ideal adaptation for each organism, using the distance

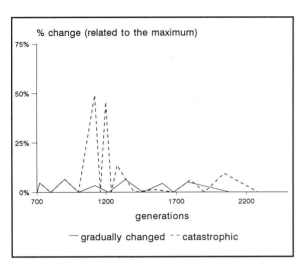

Fig. 3. The percent of the maximum feasible change vs. time in the two changing environments. Note the major sudden changes about the 1100th generation.

measure DIS divided by the constant A=27x28, that represents the maximum variability of any organism (each of 28 characters has 27 possibilities). Figure 4 shows the results of both experiments.

With small gradual environmental changes, the oscillations of adaptation values are always

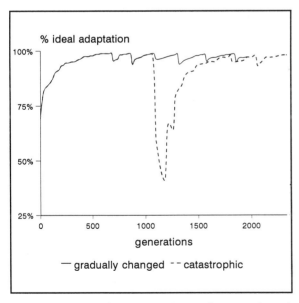

Fig. 4. The level of adaptation in two lineages through time. When adaptation is low, extinction is likely.

proportionally small. Meanwhile, in the short catastrophic period, these values come down fast and suddenly.

When the environment changes in a moderate way, the speed of adaptation does not change substantially before or after the appearance of each new environment, remaining nearly constant around its average. The situation is very different in a catastrophic period. The readaptation speed after a large environmental change increases strongly.

These results lead us to propound a general rule in our model. The potential evolutionary rate varies inversely with organisms that are poorly adapted, and vice versa. Although in our model this is not explicitly discussed, it is feasible to relate important environmental changes with mass extinctions. This simple model helps us to understand the rapid emergence of new species after catastrophies.

A sudden sharp environmental change could cause great changes in the survival of species, which could occupy vacant ecological places abandoned after extinctions (refilling the empty ecological barrel). In our model, the less adapted organisms have more potential for change, but they also have a much better chance of extinction in real life.

The Model and the Paleontologist's Work

We are aware that our simple model is far from the real work of paleontologists. They have no means of quantifying exactly the environmental characteristics, present or past, nor to establish a correlation between them and the organismal characteristics in order to measure the adaptative potential they possess at a certain moment in their evolutionary history. Due to this constraint, the usual method of paleontologists is to compare some of the organismal characteristics through time, in what they propose as an evolutionary lineage.

A classic work in this kind of research is the one made by Stanley Westoll in 1949 on morphological characteristics of fossil and living lung fish (Simpson 1983). He assigned a numerical value to the different stages of each characteristic, relative to it being primitive or specialized, and chose the values in such a way that the supposed ancestor could reach a total score of 100, and the more evolved specimen zero. These values were represented against time expressed in millions of years.

Simpson reversed these values in 1983 in order to make it possible to observe the appearance of more advanced or new characteristics, instead of the loss of ancient ones, which seemed to him a more realistic view of evolutionary change. Thus, Simpson assigned the arbitrary value of zero to the supposed ancestral specimen and the more advanced or recent of each one of the different characteristics. This treatment led him to obtain a curve very similar, but not identical, to a logistic curve (Figure 5).

To approximate our model to this method, we studied the evolutionary characteristics of our organisms compared to each other, not with their environment. We started from our sequence of time-arranged fossils and compared the characteristics (the small letters) of each organism with the ones belonging to its ancestors and its successors. The comparison is made now through the distance dij defined as:

$$DIS = |a_o - a_i| + |b_o - b_i| + |c_o - c_i| +$$

where i and j are fossils.

We take the maximum distance obtained between the most ancient phrase (we choose again the 700 generation phrase) and the newest one we call D, and give it the value of 100. Therefore, if we represent the dij/D values in percentages, our treatment is very similar to the Westoll-Simpson one. The data obtained in the experiments with gradual and catastrophic changes are represented in Figure 6.

As can be seen in the case of gradual changes,

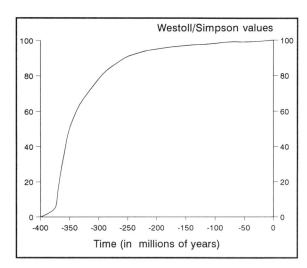

Fig. 5. Westoll's data of 16 morphological characteristics of 10 extinct and 3 extant genera of lung fish represented in Simpson's mode. The fitted curve looks like a logistic curve.

the experimental data could be adjusted to a straight line. While in the case of sharp changes, experimental data form a pseudologistical curve with a similar shape to that obtained by Westoll–Simpson.

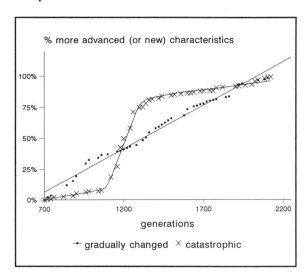

Fig. 6. Data obtained from the two evolutionary lineages in changing environments, represented in Simpson's mode. Note that the experimental points obtained from the lineage that evolved in a gradually changing environment are fitted by a straight line, whereas the data obtained from the lineage evolved in a catastrophic environment are similar to the curve of Figure 5.

We deduce from these results that, if fossil chains corresponding to lineages evolved in gradually changed environments are analyzed, the relative evolutionary rate is nearly constant along the history of life. Nevertheless, if a fossil chain includes survival ancestors from catastrophic changes in their environment, we will note punctuated periods with significant evolutionary rates inserted in lengthy periods with slow and nearly constant evolutionary rhythm.

Even if the lineage overcame catastrophic changes during the period studied, but the corresponding specimens were not found in the fossil record, the evolutionary rate studies could show gradual and small changes of rate instead of being sudden and great as expected. Incomplete fossil records can confuse things, but the content of gradualist and punctuationist theories can be studied by an analysis of these necessarily fragmented sequences.

Conclusions

Obviously, this model has a clear advantage compared with real life. It generates a complete fossil record as a single evolutionary lineage. In real life, things are different. The fossil record is very incomplete, and lineages have to be defined through complex research that includes a large number of prior considerations.

Nevertheless, in spite of the simplicity of our model and the evident differences between the present work and reality, we think that it can be useful as a teaching tool to work the content of gradualism and punctuationism theories in the classroom.

Lastly, we would like to take up again one of the aforementioned inadequacies of this model: genetic drift. This biological concept can be included in the model by changing the meaning of *organisms*. If "organism*s*" are not populations of identical individuals, as we stated first, but they play the role of different elements of the genetic space corresponding to their own species, then the several generations of "organisms" can be considered as different individuals from a population, and genetic drift is invoked, adding fitness and competition. (See Chapter 3 in Dawkins 1987, and Chapter 6 in Dobzhansky et al. 1986).

The same model could illustrate two important points of punctuated equilibrium: the rapid genetic change due to the drift and the changes of evolution rate depending on the populational adaptation capabilities, the latter one closely connected with environmental change. We think this mode could be a creative tool in practical biology teaching

Reference ·

This activity is based on an original exercise by O.B Marco and V.S. Lopez (1993). A simple model to think about the evolutionary rate in macroevolution. *The American Biology Teacher, 55*(7), 424-429, and is modified and reprinted with the permission of the publisher.

```
COMMON G. DIX, ASMO

SCREEN 2
WINDOW (-20, -20)-(120, 120)
KEY OFF: CLS
OPEN 'REG' FOR OUTPUT AS #1 'ARCHIVO 'REG' PARA
REGISTRO FOSIL

DIM F$(40)
DIM  DX(40): DIM FI(40): DIM FI(40)
DIM F(40)

50 CLS
FOR P = 1 TO 40
F(P) = 0: F$(P) = ''
FI$(P) = '': F(P) = 0
DX(P) = 0
NEXT P

60 S = 0 'NUMBER OF CHARACTERS
LOCATE 5, 1: PRINT STRING$(80, ' ') 'TO CLEAR SCREEN AREA
LOCATE 2, 1
PRINT '(USE CAPITALS) WRITE THE ATTRIBUTES OF THE
ENVIRONMENT & PRESS 'ENTER'
(MAX 30)'
PRINT ' '
PRINT STRING$(80, '*')
PRINT ' '
PRINT STRING$(80, '*')

ASMO + 0 'TOTAL ASCII VALUE (TO BE CALCULATED AFTER)

REM DEFINITION OF ATTRIBUTES OF THE ENVIRONMENT
DIXMAX = 0 'MAXIMUM POSSIBLE DISTANCE WITH ATTRIBUTES
OF THE ENVIRONMENT

FOR K = 1 TO 40

DO
F$(K) = INPUT$(1) 'CHARACTERS OF ATTRIBUTES OF THE
ENVIRONMENT
F(K) = ASC(F$(K)) 'ASCII VALUE OF CHARACTERS
IF F$(K) = CHR$(13) THEN 200 ELSE 'IF 'ENTER' THEN END
IF F(K) = 32 THEN F(K) = 64 'SPECIAL CASE: BLANK SPACE
IF F(K) > 77 THEN DM = F(K) - 64 ELSE DM = 90-64
DIXMAX = DIXMAX + DM

LOCATE 20, 5: PRINT '                              '
LOCATE 21, 5: PRINT '                              '

IF INSTR('ABCDEFGHIJKLMNOPQRSTUVWXYZ', F$(K) = 0
THEN BEEP
LOOP WHILE INSTR('ABCDEFGHIJKLMNOPQRSTUVWXYZ',
F$(K)) = 0

LOCATE 5, 5 + K  'WRITE ATTRIBUTES OF THE ENVIRONMENT
PRINT F$(K)

S = S + 1 'TO COUNT THE NUMBER OF CHARACTERS OF
ATTRIBUTES OF THE ENVIRONMENT
IF S > 30 THEN
LOCATE 20, 10
PRINT 'TOO MUCH ATTRIBUTES OF THE ENVIRONMENT'
LOCATE 21, 10
PRINT 'PLEASE, WRITE AGAIN'
GO TO 60

ELSE
END IF
ASMO = ASMO + F(K)  'TOTAL ASCII VALUE OF ATTRIBUTES OF
THE ENVIRONMENT

NEXT K  'END OF DEFINITION OF ATTRIBUTES OF THE
ENVIRONMENT

200 'LOCATE 8, 6: PRINT 'ASCII VALUE = '; ASMO 'PRINT TOTAL
ASCII VALUE OF ATTRIBUTES OF THE ENVIRONMENT

REM TO MODIFY
LOCATE 15, 3
PRINT 'TO CHANGE ATTRIBUTES, PRESS 'S'. ANOTHER KEY
TO CONTINUE'
Y$ = INPUT$(1)
IF Y$ = 's' OR Y$ = 'S' THEN 50 ELSE

CLS : G = 0

REM REWRITING ATTRIBUTES OF THE ENVIRONMENT

LOCATE 3, 20: PRINT STRING$(S + 6, '*')
LOCATE 5, 20: PRINT STRING$(S + 6, '*')
FOR J = 1 TO S
LOCATE 4, 22 + J  'PRINTS THE ATTRIBUTES OF THE
ENVIRONMENT
PRINT F$(J)
MED$ = MED$ + F$(J) 'TO RECORD TOTAL PHRASE
NEXT J
GOSUB 2000  'TO RECORD PHRASES AND DISTANCES

PRINT ' '
LOCATE 4, 1: PRINT 'ENVIRONMENT'

REM TYPING THE FIRST ANCESTOR PHRASE

LOCATE 10, 2: PRINT 'PRESS 'A' FOR FIRST ANCESTOR AT
RANDOM'
LOCATE 11, 2: PRINT 'ANOTHER KEY TO INTRODUCE THE
FIRST ANCESTOR PHRASE'

Y$ = INPUT$(1)

LOCATE 10, 2: PRINT'
LOCATE 11, 2: PRINT'

AV = 0: DIX = 0

FOR J = 1 TO S
IF Y$ = 'A' OR Y$ = 'a' THEN
X$ = TIME$; Z$ = RIGHT $(X$, 2): zm$ = MID$(X$, 4, 2): zh$
= LEFT$(X$, 2)
n = VAL(Z$) + 60 * VAL(zm$) + 3600 * VAL (zh$)
RANDOMIZE n
FI(J) = INT(RND * (91 - 64) + 64
IF FI(J) = 64 THEN FI(J) = 32
FI$(J) = CHR$(FI(J))

ELSE

LOCATE 8, 15: PRINT 'WRITE THE FIRST ANCESTOR PHRASE
OF THE LINEAGE'
DO

FI$(J) = INPUT$(1)
```

```
FI(J) = ASC(FI$(J))

LOCATE 20, 5: PRINT'
IF INSTR ('ABCDEFGHIJKLMNOPQRSTUVWXYZ'; FI$(J) = O
THEN BEEP
LOOP WHILE INSTR('ABCDEFGHIJKLMNOPQRSTUVWXYZ',
FI$(J)) = 0

END IF

IF FI(J) = 32 THEN FI(J) = 64 'FOR THE BLANK SPACE
AV = AV + FI(J)
DX(J) = ABS (FI(J) - F(J))  'ASCII DISTANCE BETWEEN THE
ATTRIBUTES OF THE ENVIRONMENT AND EVOLUTIONARY
PHRASE
DIX = DIX + DX(J)
LOCATE 10, 22 + J
PRINT FI$(J)  'PRINT EVOLUTIONARY PHRASE
FRA$ = FRA$ = FI$(J)  'TO RECORD THE ENTIRE PHRASE AS
ONE VARIABLE
NEXT J

G = 1
GOSUB 2000

PRINT ' '
LOCATE 12, 10: PRINT 'MEASURE OF ADAPTATION LEVEL:
DIS (0-100) = '; DIX

LOCATE 22, 15: PRINT 'PRESS ANY KEY TO START'
X$ = INPUT$(1)
LOCATE 22, 15: PRINT'

FOR P = 7 TO 22  'ERASE
LOCATE P, 1
PRINT STRING$(80, ' ')
NEXT P

LOCATE 7, 1: PRINT 'ORGANISMS'

600 ' START EVOLUTIONARY PROCESS
G = G+ 1
X$ = TIME$: Z$ = RIGHT$(X$, 2); zm$ = MID$(X$, 4, 2): zh$ =
LEFT$(X$, 2)
n = VAL(Z$) + 60 * VAL(zm$) + 3600 * VAL(zh$)
RANDOMIZER n

JA = INT(RND * S) + 1  'CHOOSE A PLACE IN THE EVOLUT.
ORGANISM AT RANDOM
FRA$ = ' '
FOR I = 1 TO S

IF I = JA THEN

A = INT(RND * 27) + 64  'CHARACTER AT RANDOM
D = F(I) - A
IF ABS(D) < DX(I) THEN
BAN = 1
DIX = DIX - DX(I)
AV = AV - FI(I)
DX(1) = ABS(D)
AV = AV + A
IF A = 64 THEN A = 32  'GIVING BACK THE BLANK SPACE AS 32
FI$(I) = CHR$(A)
LOCATE 8, 20: PRINT STRING$(60, ' ')
LOCATE 7, 22 + 1: PRINT FI$(I)
LOCATE 8, 22 + 1: PRINT CHR$(24)
FI(I) = ASC(FI$(I))
IF FI(I) = 32 THEN FI(I) = 64

DIX =DIX + DX(I)

ELSE
BAN = 0  'CONTROL CHANGES
END IF
ELSE

LOCATE 7, 22 + 1: PRINT FI$(I)
END IF

FRA$ = FRA$ + FI$(I)

NEXT I

LINE (-20, 60) - (110, 40) , , B
LOCATE 13, 2
PRINT 'DIS='; DIX
LOCATE 13, 50: PRINT 'GENERATION='; G

IF BAN = 1 THEN
GOSUB 2000
ELSE
END IF

IF DIX = 0 THEN 800 ELSE 600

800 REM LEE REGISTRO

GFIN = G
LOCATE 18, 1: PRINT 'MAXIMUM ADAPTATION LEVEL IN THE
GENERATION'; GFIN

LOCATE 20, 1: PRINT 'PRESS ANY KEY TO SEE THE
COMPLETE LINEAGE'
CLOSE #1

CX$ = INPUT$(1)
'WINDOW (-20, -5)-(GFIN, 150)

CLS

LOCATE 1, 1: PRINT 'PLEASE, PRESS 'P' FOR PRINTER OUTPUT
OR OTHER KEY FOR SCREEN OUTPUT'
C$ = INPUT$(1)
LOCATE 1, 1: PRINT'
OPEN 'REG' FOR INPUT AS #1
INPUT #1, MED$

IF C$ = 'P' OR C$ = 'p' THEN

LPRINT 'ATTRIBUTES OF THE ENVIRONMENT: '; MED$
LPRINT ''
LPRINT  'FIRST ANCESTOR, ASCII VALUE, DIS, & GENERATION'
ELSE
PRINT 'ATTRIBUTES OF THE ENVIRONMENT: '; MED$
PRINT ''
PRINT 'FIRST ANCESTOR, ASCII VALUE, DIS & GENERATION'
END IF

INPUT #1, FOS$, FIT, VA, ESTR
DD = 100 * (DIXMAX - FIT) / DIXMAX
IF C$ = 'p' OR C$ = 'P' THEN
LPRINT FOS$, : LPRINT VA, : LPRINT USING '###.##'; DD, :
LPRINT '   '; ESTR - GFIN
LPRINT ''
LPRINT 'FOSSIL RECORD (ORGANISMS, ASCII VALUE,
IDEAL ADAPTATION (%), & GENERATION)'
LPRINT ' '

ELSE

PRINT FOS$, : PRINT VA, : PRINT USING '###.##'; DD, :
PRINT '   '; ESTR - GFIN
PRINT ''
```

```
PRINT 'FOSSIL RECORD (ORGANISMS, ASCII VALUE, IDEAL
ADAPTATION (%), & GENERATION)'
PRINT ' '
END IF
K = 0
WHILE NOT EOF(1)
INPUT #1, FOS$, FIT, VA, ESTR
FF = 100 * (DIXMAX - FIT) / DIXMAX
IF C$ = 'P' OR C$ = 'p' THEN
LPRINT FOS$, : LPRINT VA, : LPRINT USING '###.##'; FF, :
LPRINT '   '; ESTR - GFIN
ELSE

K = K + 1
IF K < = 15 THEN
PRINT FOS$, : PRINT VA, : PRINT USING '###.##'; FF, :
PRINT '   '; ESTR - GFIN
ELSE
PRINT 'PRESS ANY KEY TO CONTINUE'
BB$ = INPUT$(1)
CLS
K = 0
END IF
END IF

WEND

CLOSE #1
PRINT ''
PRINT 'END OF LINEAGE RECORD, PRESS 'S' TO START
AGAIN OR 'E' TO EXIT'
DO
G$ = INPUT$(1)
IF INSTR('SEse', G$) = O THEN BEEP
LOOP WHILE INSTR('ES,es', G$) = 0
IF G$ = 'S' OR G$ = 's' THEN
RUN 'AZAR4'
ELSE
END IF
END

2000 REM TO RECORD PAST ORGANISMS AND DISTANCES
IF G = 0 THEN
WRITE #1, MED$
ELSE
WRITE #1, FRA$, DIX, AV, G
END IF

RETURN
```

X. GLOSSARY

Adaptation — Any characteristic that helps an organism to survive so that it may reproduce. A characteristic that may be an adaptation in one environment is not necessarily an adaptation in all environments.

Allele — One of the various forms of a gene for a particular trait. For instance, ear lobe attachment is controlled by a pair of genes with two alleles for the trait, attached or free.

Analogy — Two or more structures that have the same function, but different evolutionary origins. The classic example is the butterfly wing and the wing of a bird. Contrast with *homology*.

Apostatic Selection — Selecting common forms of prey while ignoring rare ones.

Artficial Classification — A classification scheme that uses any traits to place organism in categories with no concern for their evolutionary relationships. For instance, classifying organisms into groups by color, size, or shape would be an artifical system. This explains why whales and fish are frequently grouped together.

Biostratigraphy — Stratigraphy is the science examining the nature of rock layers (strata), and biostratigraphy is the science using the additional evidence of fossils to investigate strata.

Cladistic Classification — A method of classification in which animals and plants are placed into taxonomic groups when they share characteristics that are thought to indicate common ancestry. It is based on the assumption that two new species are formed suddenly, by splitting from a common ancestor, and not by gradual evolutionary change.

Constructivism — A learning theory suggesting that the most important predictor of future learning is what the learner already knows. This prior knowledge affects the way in which learners observe the world around them and is the foundation on which future learning is based.

Creationism — A nonscientific idea that has the central premise that species have not evolved but were created individually and independently of each other.

Cryptic Coloration — Coloration that serves to conceal, especially in animals.

Darwin-Wallace Model — The explanation of how evolution occurs -- proposed independently by Charles Darwin and Russell Wallace in the mid-1800s. (See *natural selection.*)

Descent with Modification — The term that Darwin used to describe what is now called "evolution by natural selection." (See *natural selection.*)

Evolution — The idea that the species alive today have descended (with changes) from related species that lived in the past.

Evolution/Creation Controversy — (See *creationism.*)

Gradualism — A principle inherent in the Darwin-Wallace Model of evolution by natural selection that there has been constant slow change in species through time. Contrast this with the idea of *punctuated equilibrium.*

Hardy-Weinberg Equation, Principle, or Law — A mathematical relationship seen in large randomly-mating populations. The law states that the gene frequency in the population stays the same as long as mutations, differential mating, and gene selection do not occur. The mutations causing new characteristics in organisms violate Hardy-Weinberg but provide the raw material of evolution.

Heterozygous — A condition in which the pair of genes that code for a particular trait contain the same alleles. In the case of eye color — controlled by a single pair of genes — a heterozygous condition exists when one gene codes for blue eye color, and the other gene codes for brown. Compare with *homozygous.*

Homology — Structures that now may look quite different and are descended from a common ancestral form. The arm of a human and the wings of birds and bats are homologous structures. Contrast with *analogy.*

Lamarckism — The idea proposed by Jean Baptist de Lamarck suggesting that changes in an organism during its life will affect offspring of that individual. Also known as the principle of "use and disuse." We now know that no changes in body cells will have an effect on the nature of an organism's offspring.

Macroevolution — Evolutionary change involving relatively large and complex steps.

Microevolution — Evolutionary change resulting from selective accumulation of minute variations. Contrast with *macroevolution.*

Natural Classification — A classification scheme that uses evolutionarily-derived traits to place only related organisms together in categories. For instance, although whales and dogs appear to be unrelated, they have enough evolutionarily-derived characteristics that they are grouped together in the same class.

Natural Selection — A theory used to explain *how* evolution occurs. In summary, natural selection states that there is natural variation within members of a species; species produce more offspring than can survive; some characteristics are favored over others because of environmental conditions. Those individuals favored by the environment because of the characteristics they possess will survive, reproduce and pass favored traits — and others — on to the next generation.

New Synthesis — A reference to research that has been done with respect to evolutionary biology since the time of the Darwin-Wallace model. The *New Evolutionary Synthesis* has improved our knowledge of evolution but has neither negated the theory of evolution by natural selection nor the fact of evolution itself.

Parthenogenesis — Reproduction by development of an unfertilized gamete that occurs especially among lower plants and invertebrate animals.

Piagetian Framework — A reference to the work of Jean Piaget, who has shown that individuals move through several mental stages — called developmental stages — before reaching fully-abstract thinking. Educators are advised to structure learning activities at an appropriate developmental level.

Phenetic Classification — Classificatory systems or procedures that are based on overall similarity -- usually of many characters -- without regard to the evolutionary history of the organisms involved.

Phylogeny — Line of evolutionary descent. Modern taxonomy is founded on the principle of phylogeny so that organisms that are thought to be descended from each other are classified together.

Phyletic Gradualism — (See *gradualism.*)

Punctuated Equilibrium — A new interpretation of the mode and tempo evolution proposed by Gould and Eldredge in which species remain unchanged (in equilibrium) for long periods of time, and then speciation suddenly (punctuated) occurs. Contrast this with *gradualism.*

Speciation — The separation of one ancestral species into two different species. Speciation is thought to occur when a subpopulation of the ancestral group is separated for a prolonged period and exposed to different environmental conditions.

Systematics — The study of classification systems and relationships among organisms.

Taxonomy — The study of the classification of organisms.

XI. REFERENCES CITED

Allen, J.A. (1976). Further evidence for apostatic selection by wild passerine birds. *Heredity, 36*(1), 73-180.

Allen, J.A. (1988). Frequency-dependent selection by predators. *Philosophical Transactions of the Royal Society of London,* B, 319, 485-503.

Allen, J.A. & Clarke, B.C. (1968). Evidence for apostatic selection by wild passerines. *Nature, 220*(5166), 501-502.

Allen, J.A. & Cooper, J.M. (1985). Cypsis and masquerade. *Journal of Biological Education,19*(4), 268-270.

Alters, B.J. & McComas, W.F. (1994). Punctuated equilibrium: The missing link in evolution education. *The American Biology Teacher, 56*(6), 334-340.

American Association for the Advancement of Science (AAAS). (1989). *Project 2061: Science for all Americans.* Washington, DC: Author.

Association for Science Education (ASE). (1981). *Education through science.* An ASE Policy Statement. Hatfield, UK: Author.

Attenborough, D. (1979). *Life on Earth.* Boston: Little, Brown and Company.

Bantock, C.R. & Harvey, P.H. (1974). Color polymorphism and selective predation experiments. *Journal of Biological Education, 8*(6), 323-329.

Benton, M. (1993). Four feet on the ground. In S.J. Gould (Ed.), *The book of life.* New York: W.W. Norton and Company.

Biological Science Curriculum Study (BSCS). (1992). *Evolution: Inquiries into biology and earth science.* Colorado Springs, CO: Author.

Bishop, J.E. (1992). New way to develop high–tech drugs Monkeys with Darwin's famed theory. *The Wall Street Journal,* 2/25.

Bishop, B.A. & Anderson, C.W. (1986). *Student conceptions of natural selection and its role in evolution.* E. Lansing, MI: Institute for Research on Teaching. (ERIC Document Reproduction Service No. ED 269 254)

Borror, D.J., & White, R.E. (1970). *A field guide to the insects.* Boston: Houghton Mifflin and Company.

Brownoski, J. (1973). *The ascent of man.* Boston: Little, Brown and Company.

Brumby, M.N. (1984). Misconceptions about the concept of natural selection by medical biology students. *Science Education, 68*(4), 493-503.

Burns, J.M. (1968). A simple model illustrating problems of phylogeny and classification. *Systematic Zoology, 17*(1), 170-173.

Cain, A.J. (1983a). Cepaea nemoralis and hortenis. *Biologist*, 30, 193-200.

Cain, A.J. (1983b). Ecology and ecogenetic of terrestrial molluscan populations. In R. Russel–Hunter (Ed.), *The Mollusca 6,* (pp. 597-647). London: Academic Press.

Cain, A.J. & Sheppard, P.M. (1954). Natural selection in Cepaea. *Genetics*, 39, 89–116.

Campbell, N. (1953). *What is science?* New York: Dover.

Carey, R.L. & Strauss, N.G. (1970). An Analysis of experienced science teachers' understanding of the nature of science. *School Science and Mathematics, 70*(5), 366-376.

Cesnola, A.P. di (1904). Preliminary note on the protective value of color in *Mantis religiosa. Biometrica, 3*(1), 58-59.

Chalmers, A.F. (1982). *What is this thing called science?* 2nd ed., Queensland: University of Queensland Press.

Clarke, B.C. (1962). Balanced polymorphism and the diversity of sympatric species. In D. Nicholas (Ed.), *Taxononomy and geography* (pp. 47-70). Oxford: Systematics Association.

Clarke, B.C., Arthur, W., Horsley, D.T. & Parkin, D.T. (1978). Genetic variation and natural selection in pulmonate molluscs. In V. Fretter & J. Peake (Eds.), *Pulmonates, Vol. 2A: Systematics, evolution and ecology* (pp. 219-270). New York: Academic Press.

Clough, M.P. (1989). *Recommendations for preservice and inservice preparation of teachers in the nature of science.* Position statement submitted to the National Association of Biology Teachers' Teacher Education Committee.

Cooper, J.M. (1984). Apostatic selection on prey that match the background. *Biological Journal of the Linnean Society, 23*(1), 221-228.

Custer, T. (1971). *Breeding biology of the Alaskan longspur.* Unpublished master's thesis, California State College, Fullerton, California.

Darwin, C. (1975). *The origin of species.* Introduced and abridged by Philip Appleman. New York: W.W. Norton.

Dawkins, M. (1971). Perceptual changes in chicks: Another look at the 'searching image' concept. *Animal Behaviour, 19*(2), 566-574.

Dawkins, R. (1987). *The blind watchmaker.* New York: W.W. Norton.

Dixon, D. (1981). *After man.* London: Harrow House Editions, Ltd.

Dobzhansky, T. (1973). Nothing in biology makes sense except in the light of evolution. *The American Biology Teacher, 35*(3), 125-129

Dobzhansky, T., Ayala, F.J., Stebbins, G.L. & Valentine, J.W. (1986). *Evolution.* San Francisco, W.H. Freeman.

Dobzhansky, T. & Epling, C. (1944). *Contributions to the genetics, taxonomy and ecology of Drosophila pseudoobscura and its relatives.* Washington, DC: Carnegie Institution.

Edmunds, M. (1974). *Defense in animals.* Harlow: Longman.

Einstein, A. & Infeld, L. (1938). *The evolution of physics.* New York: Simon and Schuster.

Eldredge, N. & Gould, S.J. (1972). Punctuated equilibria: An alternative to phyletic gradualism. In T.J. Schopf (Ed.), *Models in paleobiology* (pp. 82-115). San Francisco: Freeman, Cooper.

Eldredge, N. (1989). *Time frames: The evolution of punctuated equilibria.* Princeton: Princeton University Press.

Endler, J. A. (1981). An overview of the relationships between mimicry and crypsis. *Biological Journal of the Linnean Society,16*(1), 26-31.

Etkin, W. (1967). *Social behavior from fish to man.* Chicago: University of Chicago Press.

Eve, R.A. & Dunn, D. (1990). Psychic powers, astrology and creationism in the classroom? *The American Biology Teacher, 52*(1), 10-21.

Futuyma, D.J. (1986). *Evolutionary biology.* Sunderland: Sinauer Associates, Inc.

Gallagher, J.J. (1984). Educating high school teachers to instruct effectively in science and technology. In R. Bybee, J. Carlson & A.J. McCormack (Eds.), *Redesigning Science and Technology Education – NSTA Yearbook 1984.* Washington, DC: National Science Teachers Association.

Gould, S.J. (1977). *Darwin's untimely burial. Ever since Darwin.* New York: Norton.

Gould, S.J. (1977). Evolution's erratic pace. *Natural History, 86*(5), 12-16.

Gould, S.J. (1987). Justice Scalia's misunderstanding. *Natural History, 96*(10), 14-21.

Gould, S.J. (1991). Opus 200. *Natural History, 100*(8), 12-18.

Gould, S.J. (1988). The heart of terminology: What has an abstruse debate over evolutionary logic got to do with Baby Fae? *Natural History, 97*(2), 24-31.

Gould, S.J. (1989). *Wonderful life. The Burgess shale and the nature of history.* New York: W.W. Norton.

Gould, S.J. & Eldredge, N. (1977). Punctuated equilibria: The tempo and mode of evolution reconsidered. *Paleobiology, 3*(2), 115-151.

Greenwood, J.J.D. (1984). The functional basis of frequency–dependent food selection. *Biological Journal of the Linnean Society, 2/3,* 177-199.

Greenwood, J.J.D. (1985). Frequency–dependent selection by seed predators. *Oikos, 44*(1), 195-210.

Greenwood, J.J.D. & Elton, R.A. (1979). Analyzing experiments on a frequency–dependent selection by predators. *Journal of Animal Ecology, 48*(3), 721-737.

Greenwood, J.D.D., Wood, E.M. & Batchelor, S. (1981). Apostatic selection of distasteful prey. *Heredity, 57*(1), 27-43.

Harris, S. (1977). *What's so funny about secience?* Los Altos, CA: Kaufmann.

Higgins, R.C. (1974). Specific status of *Echinocardium cordatum, E. australe* and *E. zealandicum* around New Zealand. *Journal of Zoology, 173*(4), 451-75.

Hodson, D. (1988). Toward a philosophically more valid science curriculum. *Science Education, 72*(1), 19-40.

Hubbard, S.F. Cook, R.M., Glover, J.G. & Greenwood, J.J.D. (1982). Apostatic selection as an optimal foraging strategy. *Journal of Animal Ecology, 51*(2), 625-633.

Huxley, J. (1966). *The Galapagos.* Berkeley, CA: University of California Press.

Jones, J.S., Leith, B.H. & Rawlings, P. (1977). Polymorphism in Cepaea: a problem with too many solutions? *Annual Review of Ecology and Systematics, 8,* 109-143.

Johnson, R.L. & Peeples, E.E. (1987). The role of scientific understanding in college: Student acceptance of evolution. *The American Biology Teacher, 49*(2), 93-98.

Keown, D. (1988). Teaching evolution: Improved approaches for unprepared students. *The American Biology Teacher, 50*(7), 407-410.

Kettlewell, H.B.D. (1955) Selection experiments on industrial mechanism in the Lepidoptera. *Heredity, 9*(3), 323-342.

Kettlewell, H.B.D. (1956) Further selection experiments on industrial melanism in the Lepidoptera. *Heredity, 10*(3), 287-301.

Kettlewell, H.B.D. (1959). Darwin's missing evidence. *Scientific American, 200*(4), 48-53.

Kettlewell, H.B.D. (1973). *The evolution of melanism: The study of a recurring necessity.* Oxford: Oxford University Press.

Kitcher, P. (1982). *Abusing science: The case against creationism.* Cambridge, MA: The MIT Press.

Koestler, A. (1971). *The case of the midwife toad.* London: Hutchinson.

Kuhn, T.S. (1970). *The structure of scientific revolutions.* Chicago: University of Chicago Press.

Lack, D. 1953. Darwin's finches. *Scientific American, 188*(4), 66-72.

Lawson, A.E. & Worsnop, W.A. (1992). Learning about evolution and rejecting a belief in special creation: Effects of reflective reasoning skill, prior knowledge, prior belief and religious commitment. *Journal of Research in Science Teaching, 29*(2), 143-166.

Lewin, R. (1980). Evolutionary theory under fire. *Science, 210*(21), 883-887.

Manuel, D.E. (1981). Reflections on the role of history and philosophy of science in school science education. *School Science Review, 62*(221), 769-771.

Matthews, M.R. (1989). A role for history and philosophy in science teaching. *Interchange, 20*(2), 3-15.

Mayr, E. (1982). *The growth of biological thought: Diversity, evolution, and inheritance.* Cambridge, MA: The Belknap Press of Harvard University Press.

Moment, G.B. (1962). Reflexive selection: A possible answer to an old puzzle. *Science,* 136, 262-263.

Moore, J.A. (1993). *Science as a way of knowing: The foundations of modern biology.* Cambridge: Harvard University Press.

Moore, J.A. (1983). Evolution, education, and the nature of science and scientific inquiry. In J. Peter Zetterberg (Ed.), *Evolution versus creationism: The public education controversy.* Phoenix, AZ: Oryx Press.

Minitzer, J.J. & Arnaudin, M.W. (1984). *Children's biology: A review of research on conceptual development in the life sciences.* (ERIC Document Reproduction Services No. ED 249 044)

Murdoch, W.W. (1969). Switching in general predators: Experiments on predator specificity and stability of prey populations. *Ecological Monographs, 39,* 335-354.

Murdoch, W.W., Avery, S. & Smyth, M.E.B. (1975). Switching in a predatory fish. *Ecology,* 56, 1094-1105.

National Assessment of Education Progress (NAEP). (1989). *Science Objectives,* Princeton: Author.

National Science Teachers Association. (1982a). *The laboratory. An NSTA position statement.* Washington, DC: Author.

National Science Teachers Association. (1982b). *Science–technology–society: Science education for the 1980's.* An NSTA Position Statement. Washington, DC: Author.

Newcomb, E.H., Gerloff, G.C. & Whittingham, W.F. (1964). The selective role of the environment in evolution. *Laboratory studies in biology.* San Francisco: W.H. Freeman and Company.

Nunan, E. (1977). History and philosophy of science and science teaching: A revisit. *The Australian Science Teachers Journal, 23*(2), 65-71.

Otto, J.H. & Towle, A. (1973). *Modern biology.* New York: Holt, Rinehart and Winston.

Patterson, J.T. & Wheeler, M.R. (1942). *Description of new species of the subgenus Hirtodrosophila and Drosophila.* Austin, TX: University of Texas Press.

Popper, K.R. (1963). *Conjectures and refutations: The growth of scientific knowledge.* New York: Harper & Row.

Prakash, S. (1977). Gene polymorphism in natural populations of *Drosophila persimilis. Genetics, 85*(3), 513-520

Reader, J. (1986). *The rise of life: The first 3.5 billion years.* New York: Knopf.

Renner, J.W. (1981). Why are there no dinosaurs in Oklahoma? *The Science Teacher, 48*(9), 22-25.

Rhodes, F.H.T. (1984). Science correspondence: Darwin's gradualism and empiricism. *Nature, 30*(5964), 116.

Ridley, M. (1985). *The problems of evolution.* Oxford: Oxford University Press.

Rosenthal, D.B. (1979). Using species of *Drosophila* to teach evolution. *The American Biology Teacher, 41*(9), 552-555.

Rowe, R.E. (1976). *Conceptualizations of the nature of scientific laws and theories held by middle school and junior high school teachers in Wisconsin.* Unpublished doctoral thesis. The University of Wisconsin -Madison.

Rowell, J.A. & Cawthorne, E.R. (1982). Image of science: An empirical study. *European Journal of Science Education, 4*(1), 79-94.

Rubba, P.A., Horner, J.K. & Smith, J.M. (1981). A study of two misconceptions about the nature of science among junior high school students. *School Science and Mathematics, 81*(3), 221-226.

Ryan, A.G. & Aikenhead, G.S. (1992). Students' preconceptions about the epistemology of science. *Science Education, 76*(6), 559-580.

Sagan, C. (1980). *Cosmos.* New York: Random House.

Saunders, W.L. (1992). The constructivist perspective: Implications and teaching strategies for science. *School Science and Mathematics, 92*(3), 136-141.

Scharmann, L.C. & Harris, W.M. (1992). Teaching evolution: Understanding and applying the nature of science. *Journal of Research in Science Teaching, 29*(4), 375-388.

Sheppard, P.M. (1951). Fluctuations in the selective value of certain phenotypes in the polymorphic land snail *Cepaea nemoralis. Heredity, 5*(1), 125-34.

Simpson, G.G. (1983). *Fossils and history of life.* San Francisco: Scientific American Books.

Sokal, R.R. (1966). Numerical taxonomy. *Scientific American, 215*(6), 106-116.

Stewart, I. (1990). Representación matematizada de las species de sus aptitudes y del curso de su evolución. *Investigación y Ciencia, 168*(1), 85-91.

Strurtevant, A.H. (1921). *The North American species of Drosophila.* Carnegie Institution of Washington Publication 301. Washington, DC: Carnegie Institution.

Summers, M.K. (1982). Philosophy of science in the science teacher education curriculum. *European Journal of Science Education, 4*(1), 19-27.

Tordoff, W. (1980). Selective predation of gray jays, *Perisoreus canadenis,* upon boreal chorus frogs, *Pseudaeris triseriata. Evolution, 34*(5), 1004-1008.

Van Andel, T. (1981). Consider the incompleteness of the fossil record. *Nature, 294*(5840), 397-398.

Washburn, S.L. & Devore, I. (1961). The social life of baboons. *Scientific American, 204*(6), 62-71.

Webb, S.D. (1968). *Evolution: A laboratory block.* Boston: D.C. Heath and Company.

Whitfield, P. (1993). *From so simple a beginning: The book of life.* pp. 178-181, New York: Macmillan Publishing Company.

Wilson, E. O. (1992). *The diversity of life.* pp. 88-89, Cambridge, MA: The Belknap Press of Harvard University Press.

XII. Acknowledgments

Sincere appreciation is extended to the following organizations and individuals for their cooperation in granting permission for articles in their publications to be modified and reprinted in this monograph:

The American Biology Teacher
Patricia J. McWethy, Executive Director
National Association of Biology Teachers
11250 Roger Bacon Drive, #19
Reston, Virginia 22090

Journal of Biological Education
Jonathan Cowie, Publications Manager
Institute of Biology
20-22 Queensbury Place
London, UK SW7 2DZ

Journal of College Science Teaching
Michael Byrnes, Managing Editor
National Science Teachers Association
3140 North Washington Boulevard
Arlington, Virginia 22201

Journal of Geological Education
Jim Shea, Editor
Appalachian State University
Department of Geology
Boone, North Carolina 28608

School Science Review
Jane R. Hanrott, Assistant Secretary
Association for Science Education
College Lane, Hatfield
Hertfordshire, UK AL 10 9AA

Systematic Zoology
David M. Hillis, President
Society of Systematic Zoologists
University of Texas
Department of Zoology
Austin, Texas 78712